UNIVERSITY OF STRATHCLYDE

30125 00640919 6

KV-038-692

ANDERSONIAN LIBRARY
★
WITHDRAWN
FROM
LIBRARY
STOCK
★
UNIVERSITY OF STRATHCLYDE

Books are to be returned on or before
the last date below.

- 9 JAN 2001

LIBREX —

Education in Automotive Engineering

Co-sponsoring Organizations

IMechE
Conference Transactions

International Conference on

Education in Automotive Engineering

11 November 1999
National Exhibition Centre, Birmingham, UK

Organized by
The Automobile Division of the
Institution of Mechanical Engineers (IMechE)

and

Centre Exhibitions
A member of The NEC Group

With special thanks to:

IMechE Conference Transactions 1999–10

**Professional
Engineering
Publishing**

Published by Professional Engineering Publishing Limited for The Institution of
Mechanical Engineers, Bury St Edmunds and London, UK.

First Published 1999

This publication is copyright under the Berne Convention and the International Copyright Convention. All rights reserved. Apart from any fair dealing for the purpose of private study, research, criticism or review, as permitted under the Copyright, Designs and Patents Act, 1988, no part may be reproduced, stored in a retrieval system, or transmitted in any form or by any means, electronic, electrical, chemical, mechanical, photocopying, recording or otherwise, without the prior permission of the copyright owners. *Unlicensed multiple copying of the contents of this publication is illegal.* Inquiries should be addressed to: The Publishing Editor, Professional Engineering Publishing Limited, Northgate Avenue, Bury St Edmunds, Suffolk, IP32 6BW, UK. Fax: +44 (0)1284 705271.

© 1999 Institution of Mechanical Engineers, unless otherwise stated

ISSN 1356–1448
ISBN 1 86058 225 7

A CIP catalogue record for this book is available from the British Library.

Printed by The Book Company, Ipswich, Suffolk, UK.

The Publishers are not responsible for any statement made in this publication. Data, discussion, and conclusions developed by authors are for information only and are not intended for use without independent substantiating investigation on the part of potential users. Opinions expressed are those of the Author and are not necessarily those of the Institution of Mechanical Engineers or its Publishers.

Automotive Industry Event Steering Committee

C Ashley
Steering Committee Chairman

C Bale
Facilitator

A Cole
University of Central England

T Crisp
Jaguar Cars Limited

C Ennos
Consultant

M Gidlow
Automobile Division Chairman

R Johnson
MSX International

P Jones
SMMT

A Roberts
DTI

B Watson
SMMT

P Willmer
SMMT Design Group

Conference Organizing Committee

A Cole (Chairman)
University of Central England

D Crolla
University of Leeds

K Mortimer
Consultant

G Rogers
University of Central England

Related Titles of Interest

Title	Editor/Author	ISBN
IMechE Engineers' Data Book	Clifford Matthews	1 86058 175 7
Advances in Manufacturing Technology – XIII (NCMR 99)	A N Bramley, A R Mileham, L B Newnes, and G W Owen	1 86058 227 3
The Continuum of Design Education (EDE99)	N P Juster	1 86058 208 7
Design Reuse – Engineering Design Conference '98	Dr S Sivaloganathan and Dr T M M Shahin	1 86058 132 3
Managing Enterprises – Stakeholders, Engineering, Logistics, and Achievement	D T Wright, M M Rudolph, V Hanna, D Gillingwater, and N D Burns	1 86058 066 1

For the full range of titles published by Professional Engineering Publishing contact:

Sales Department
Professional Engineering Publishing Limited
Northgate Avenue
Bury St Edmunds
Suffolk
IP32 6BW
UK

Tel: +44 (0)1284 724384
Fax: +44 (0)1284 718692

At Visteon, we're doing things differently.

The fact is, with nearly 100 years of automotive experience behind us, we already have 81 manufacturing facilities on five continents. Over 2000 patents issued and pending worldwide. With 82,000 employees, we are one of the largest automotive suppliers, working to integrate 27 areas of systems technology.

Even better, we have a brand new vision of how to put it all to work to create innovative, enterprising and individual solutions for you. Just like we already have for Daewoo, Fiat, Ford, General Motors, Jaguar, Mercedes, Toyota and The VW Group.

If, like them, you prefer to do business with an organisation committed to your success, just visit our website at **www.visteon.com**

Catch the new wave of thinking in integrated automotive systems.

See the possibilities™

Visteon

Contents

Partnership with Industry

C574/008/99	NVH education in industry and academia M F Russell	3
C574/028/99	Sustainable learning in the automotive supply chain N Barlow, A C Lyons, P F Chatterton, A Glover, M Jones, and B Oxtoby	21

Undergraduate

C574/031/99	A design-and-build racing car project: the changing face of automotive engineering at the University of Leeds A J Deakin, P C Brooks, M Priest, D C Barton, and D A Crolla	33

Postgraduate

C574/019/99	Developing engineers in the automotive industry P R Bullen, P B Taylor, and H Mughal	45
C574/024/99	The role of research in learning and personal development for engineering excellence in the automotive industry A J Day, R S F Harding, and K W Mortimer	55
C574/037/99	Development of a Master of Science programme in automotive systems engineering S J Walsh, A Malalasekera, and T J Gordon	67
C574/021/99	The virtual automotive learning environment P R Bullen, F Haddleton, P B Taylor, M Young, and P F Chatterton	77

Skills and Competencies

C574/011/99	Critical competencies for automotive engineers – how to identify and develop the necessary skills P B Taylor, P R Bullen, and J A Mulryan	91
C574/023/99	Developing student capabilities and improving the local skills base – a new venture in work-based learning with the automotive industry I Dunn, B Porter, and R Perks	101

Industrial PhD

C574/012/99 **Global Product Development – an integral part of an engineering doctorate in automotive engineering management**
P B Taylor, P R Bullen, and M D Cook 113

Authors' Index 121

Partnership with Industry

C574/008/99

NVH education in industry and academia

M F RUSSELL
Diesel Systems, Lucas Varity, Gillingham, UK

1   SYNOPSIS

This paper compares the different experience of postgraduate education and training in industry and in a university department. Post experience courses for engineers in industry need careful course design and a clear structure in physical principles to provide a framework for rapid and accurate assimilation by engineers. In contrast, students reading for an M Sc. in a UK University prefer a mathematical derivation structure, but need practical examples to survive in industry. Case studies of practical applications of these principles are discussed in the different environments and some assessment criteria are suggested for industrial and academic courses.

2   INTRODUCTION

From time to time, industrial concerns find that their prime resource has become less effective because it is not using the best available knowledge and most efficient processes in the Company's operations. Responses to this revelation varies from recruiting people who are perceived to have better understanding of the technology, through sending their key people on courses, to providing training in house. Unless the real requirement has been analysed properly, all these courses of action can fail. Common problems are: -

- Expensive transplants fail to adapt to the organisation which recruited them.
- Student and graduate training programmes produce well-trained engineers who leave the company shortly after they perceive their training to be complete.
- Sending staff on MSc courses run by universities provides them with the opportunity to find a new job, particularly if the organisation fails to recognise their enhanced worth.
- Sending staff out to short courses fails to transfer much useable knowledge into the company unless they have to use, consolidate and extend this knowledge on their return.
- In-house courses generate enthusiasm in the short term but no lasting gain.

This paper sets out what the diesel fuel injection industry requires of engineers and discusses some case studies of training and education projects.

3   DEFINING THE REQUIREMENTS FOR TRAINING AND EDUCATION

The core requirement for professional engineers in manufacturing industry is that they can create new products to fill a market requirement and provide them in sufficient volume and at a cost which satisfies the market. From experience, the overall process includes: -
- Identifying a market requirement.
- Writing a detailed engineering specification for a new product or the adaptation of an existing product which will meet this requirement in all foreseeable service conditions.
- Creating a design to meet this requirement, including: reading, brainstorming and innovating drawing, numerically assessing performance at the concept stage, costing, "benchmarking" alternative design concepts, analysing the stresses in critical components, selecting bought-in components, detailing and specifying tolerances.
- Building models of product performance and validating them with experimental results.
- Building and testing prototypes; interpreting the results and incorporating improvements.
- Documenting the specification, design, performance, any problems and actions to fix them.
- Redesigning the proven concept for ease of manufacture and ease of assembly.
- Presenting the case for investment with salient facts in a logical sequence to decision-makers.
- Championing his, or her, design when the investment in manufacturing facilities is discussed.
- Devising cost-effective manufacturing methods and specifying the plant and equipment needed.
- Purchasing and commissioning the plant and equipment needed.
- Establishing efficient and environmentally friendly manufacturing processes and practices.
- Reducing the cost of manufacture by continually introducing improvements, large and small.
- Meeting the demand for the product on time and with a quality which satisfies the customers.
- Providing after-sales service.

Another approach to describing the range of tasks undertaken by engineers is to examine the range of responses to customers' requests; as illustrated in Fig. 1 for a fuel injection manufacturer.

In developing products or processes, problems will arise which need to be defined precisely, analysed and solved. Alternative solutions may be assessed by modelling; but the fastest route is usually a combination of simulation and testing. Many experienced engineers find job satisfaction in design and development departments where they can exercise their analytical skills and the engineering judgement they have learned from product development projects.

In small companies, the range of responses is likely to be less varied than those in Figure 1; in particular the major product changes shown to the right hand side of the figure are mercifully rare. However individual engineers may become involved in many more activities. In large organisations engineers often have more opportunity to develop specialist skills and expertise.

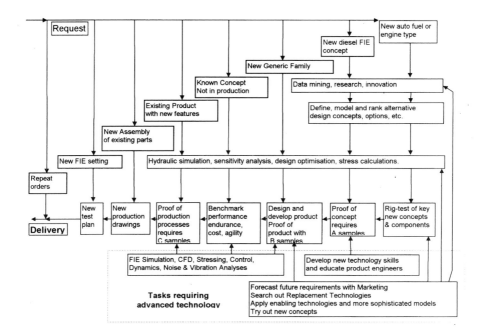

Figure 1 Responses from the engineering departments of a fuel injection manufacturer to customer requests covering the range from repeat orders to a new engine concept. Prototypes and subsequent samples are classified into: - A-samples which are made to prove the concept, B-samples to prove the design and C-samples to prove the production processes.

The various processes in Figure 1 require different skills, procedures and knowledge. To rise to a prominent position, the engineer needs to experience a fair proportion of these. To this end, he, or she, needs to learn rapidly and adapt to new teams to succeed at a variety of tasks. In the process he, or she, will learn additional skills such as project management, FMEA, value engineering, statistical analysis techniques, human resource management and how to run production processes. By working in each function engineers become familiar with the detail, calculation procedures and processes which together form the company technology base.

The demand for more powerful, refined and lower cost products and components in the automotive industry and "technology push" combine to require continuing need to learn new skills and better operational procedures of all sorts. Figure 2 summarises the innovation process for a new fuel injection product. The use of well-validated simulations early in the process is crucial to successful product innovation and development.

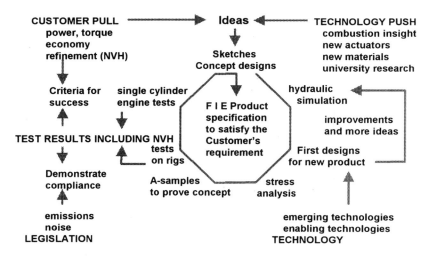

Figure 2 Steps to create a new fuel injection product to proof of concept stage. Further iterations, or similar processes are needed to design for volume production and ease of assembly.

The measurement, interpretation and reduction of noise and vibration is a particularly specialist role in the automotive and other industries. To acquire experimental data, engineers need some specialist experimental skills and a clear understanding of the acoustic and vibration behaviour of the product, plant or phenomena to be measured. Some skill in mathematics and statistics is required to reduce experimental data to "results" suitable for interpretation. The engineering expertise developed by hard experience probably adds most value in putting new results into context and interpreting them. Further skill in mathematics is required to construct accurate and useful models of noise and vibration phenomena. For someone who is highly motivated to specialise in the subject, the personal investment in a dedicated first degree or postgraduate course will pay dividends. For those who start their industrial career in a noise or vibration job, the understanding of dynamic phenomena and the experimental and theoretical skills needed are good preparation for other roles.

Alternatively, an engineering career may include a period in noise or vibration as well as several other departments; and the ability to pick up the essentials of the subject quickly and become an effective team member is more important. For this career pattern the specialist courses may be too large an investment but experience suggests that the individual noise or vibration courses in a general engineering degree seem to be inadequate. Candidates for NVH jobs with this education rarely succeed at the technical interview unless they have worked in a noise or vibration job previously, or they have better than average mathematical skills and can apply them. A good Mechanical Engineering education plus post-experience training and education in noise and vibration would seem to be more appropriate for such a career pattern.

The requirement for post-experience training has been evident for a long time. The first two case studies arose from a need to train and educate qualified engineers from several branches of the profession to tackle noise problems in production facilities. The second two case studies discuss the closest comparable modules which formed part of a course leading to an MSc degree.

4 CASE STUDY 1 NOISE MEASUREMENT TRAINING

4.1 Training Requirement
This training requirement arose from a Company directive to find cost-effective ways to tackle rising noise levels in the Company's production facilities. A task force of senior engineers with experience in noise problems, medical officers and purchasing officers drew up a comprehensive strategy to tackle noise problems[1]. Maximum permissible sound levels were set and specified, with appropriate and simple measurement procedures, in Company Standards.

The Company was purchasing up to 1000 major items of plant and equipment per year, so the first task was to ensure that all new plant and equipment complied with an appropriate noise standard. As a new feature of the existing acceptance test procedure, the noise from each machine was to be checked at the supplier's premises. The cost of setting up a central specialist group to undertake this work was prohibitive. The decision was taken to train at least 100 Methods Engineers, Machine Tool Examiners and Production Engineers to provide reliable noise measurements upon which purchasing decisions would be made. Plant suppliers were expected to challenge some measurements so a small central team of Noise Control Engineers was set up to provide support and authoritative recommendations on cost-effective control.

4.2 The Course Participants
The company had specifications for each job, so the level of the course could be determined by studying these for Machine Tool Examiners, Methods and Production Engineers. The course had to attract 10 to 20 participants at a time to be cost-effective itself, so the course design was based on the premise that managers would release their scarce human resource for one day willingly, 2-3 days if pressed and for 4 or more days not at all.

4.3 Course Design
A one-day course was designed from the bottom up, starting from a clearly defined requirement. The material had to be delivered succinctly; which involved finding ways to express the fundamentals, concepts and results in drawings and charts, using language and a structure which would be familiar and readily accessible to the participants. Duplicated notes were provided which laid out the essentials and most of the charts, tables and sketches; the participants added to these.

The 20-minute introduction explained the Company strategy to and sought to motivate the participants to concentrate during the course. Of several alternatives tried, a tape recording of words replayed with and without a filter which replicated the AAOO threshold for hearing impairment to within ±0.5 dB proved to be most useful. The procedure embodied a challenge. In the first section, similar words were replayed first unfiltered then filtered and the participants barely took in the message. In the second section, a sequence of words was replayed with the filter in use; and the participants were asked to repeat the words. Since the speech was clearly audible with the filter, but tantalisingly unintelligible, the participants began to concentrate. The third section was a short, lively piece of music, with plenty of percussion: the filter removed much of the life from the music.

The effects of hearing impairment upon the ability to communicate were discussed with the participants; and any participant who had a relative or friend who was hard of hearing was invited to comment. Since a large proportion of the population suffer from some hearing impairment in old age, one or more participants could be expected to have experienced difficulties in talking to an elderly relative or friend; and be prepared to speak about them.

The introduction continued with a very brief description of the human ear and the nature of the damage caused by excessive exposure to high levels of noise. The Company policy and strategy to reduce noise at source by engineering modifications to the machines wherever practicable was discussed and the participants were invited to "buy in" to this strategy.

The course material was delivered in short sections, separated by demonstrations which involved the course participants. The sound level scale in decibels referred to 20 µPa was illustrated with 5 dB, 10 dB and 20 dB reductions of a noise recording from a blanking press which was quite well known in the Company. The A-weighting was illustrated by playing the same press noise through an A-weighting filter and comparing it to the unfiltered noise. This provided some audible experience to support the measurements in dB(A) re20 µPa. Octave band-pass filters were demonstrated in the same way and the participants discovered for themselves how noise sources which could be separated even with rather coarse filters.

Noise measurement equipment was described from microphone details to true RMS indicating meters. The precision of different grades of Sound Level Meters was discussed briefly.

In the last 30 minutes of the morning session, a machine (a portable drill in a stand) was set up in the lecture room and noise measurements were demonstrated. After three and a half hours, the participants were tiring and few remembered much of this demonstration.

During the afternoon the course participants made 3 sets of measurements in carefully chosen sites in a Company factory. These measurements were not just exercises, the survey results and machine noise levels were recorded in the log for the factory; providing an incentive to measure and record the results carefully and demonstrating the trust the Company was placing in the participants. The site Environmental, Health and Safety Engineer would come along and collate their results for some courses. At a fourth location, a specialist engineer demonstrated additional measurements as an example of the support from the Noise Centre.

4.4 Effectiveness of the Training

The really effective part of this course was the practical session in the afternoon. The fact that the participants were taking the actual measurements for that part of the factory was a great stimulant to concentration and commitment. Years afterwards, course participants remembered the course and the central messages, and still refer to it when met by chance.

No formal examination was made of the material learned; however, the participants were encouraged to send their results to the Noise Centre. The standard of work was excellent. Any issues which emerged were dealt with by further explanation during the next visit to the factory at which the participant was based; and the course material was improved.

In over two decades the trained engineers made almost all the noise measurements of new plant and equipment. The measurements made by these engineers were a crucial part of the Company strategy to reduce noise at its origins and their commitment made the strategy work. As plant and equipment were replaced and as existing noise problems were tackled, the levels in production facilities were reduced until operators no longer needed hearing protection. The suppliers challenged some of the measurements. However in subsequent discussions, they would comment that they "could talk to your engineers about the (noise) measurements".

5 CASE STUDY 2 NOISE CONTROL EDUCATION IN INDUSTRY

In response to a strong demand for more information on noise control techniques, a second one-day course was devised to present techniques to reduce machinery noise at source. The objective was to provide participants with sufficient understanding of noise control technology for production plant for them to be able to choose cost-effective solutions to the noise problems they are likely to encounter.

Over 500 people attended 6 courses in the first 12 months. Subsequently the course was run for groups of between 20 and 120 participants as required; until over 3000 have attended in total. Approximately half of these were from outside the Company.

5.1 Noise Control Course Design

The course design process started with a list of the problems which participants would be called upon to solve, mostly from the noise measurements of new plant. A few problems were deemed to be too difficult to tackle with only one day training, for example self-excited vibration leading to excessive noise. Work in the Noise Centre to reduce noise from diesel engines provided a basis for the course structure; Figure 3 shows the noise generating processes of a diesel engine[2,3], which provided the inspiration for the course structure.

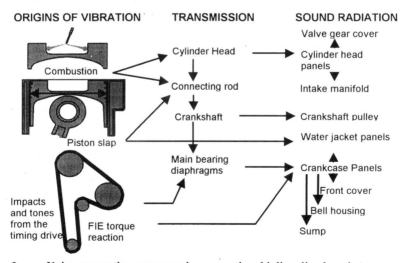

Figure 3   Noise-generating processes in conventional inline diesel engines

After some brainstorming sessions with practising Noise Control Engineers, the list of noise problems was reduced to essential machine elements which seemed to be sufficiently generic:-

5.2 Noise originating as vibration within machinery
Originating mechanisms: -
> Impacts: Blanking tools, end stops,
> Rotating Machines: Out of balance, Rolling element bearings, Cams, Gears, Chains, Hydraulic pumps & motors, Electrical machines, Toothed belt drives
> Machining and forming processes: Turning/shaping, Milling, Grinding,
> Machine Frames
> H-frame presses
> Lathes and vertical borers
> Milling machines and borers

5.3 Noise radiating surfaces (Acoustic sources)
> The workpiece itself
> Thin panels integral with the frames
> Thin cast covers
> Sheet metal panels

5.4 Noise Originating from Aerodynamic Sources
Fans
Axial fans, Centrifugal fans, Rotor windage
Air motion and turbulence
Air jets and air guns
Air flow through ducts
Airflow around teeth of toothed belt drives
Machining processes
Wood planers

5.5 Acoustic Resonances
Reverberation and Standing waves
Helmholz resonators

5.6 Palliatives
Acoustic enclosures
Acoustic duct silencers

The Noise Generating Processes chart expanded and became more detailed as the courses progressed and the range of noise problems became better defined. This chart illustrated the main part of the course structure in which course participants were taught how to break down a noise problem into categories, so that the noise control measures appropriate to each category can be selected, adapted and developed. This process had to be sufficiently flexible to find all the practical alternatives for each category, so that the participants could combine cost-effective combinations of treatments to control the noise from most plant.

The next section described two simple techniques to identify and rank noise-originating mechanisms and two more to rank noise-radiating surfaces.[4] This section introduces some

A diagram was produced to illustrate the course structure; Figure 4 is the simplest form.

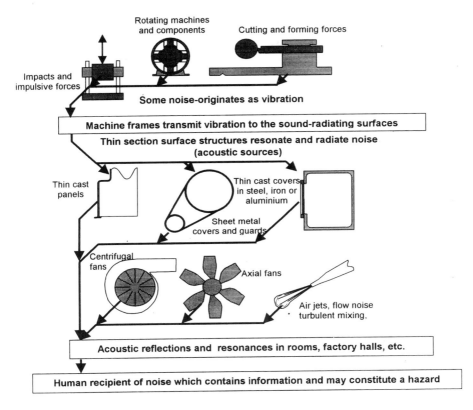

Figure 4 Noise Generation Processes chart, which formed the basis of the Noise Control Courses

new concepts to the participants; so to illustrate their use, and to provide real examples for the following sections, participants were asked to describe the noise sources on machines, which they knew well. These were categorised as a (large and noisy) group activity and the participant concerned was asked for his, or her, best engineering judgement on the ranking. Often, several participants provided joint rankings, which were based on detailed observations over several years with different tooling, materials, etc. These were reliable enough to practise the categorisation (diagnosis) process; and they were referred to throughout the rest of the course.

The course needed to describe and encourage the use of a variety of noise control treatments. These were linked to the Noise Generation Chart and the diagnosis module throughout the day.

A brief guide to the design of vibration isolation followed this with a choice from a variety of examples. Participants could design simple systems from this material, with the aid of a supplier's catalogue. Vibration damping for large structures was described. Damping for sheet metal was illustrated with a practical demonstration of real components impacting chutes.

The second most rewarding area to tackle noise is at the radiating surfaces (acoustic sources). Machine structures often benefit from stiffness; and extra mass is not such a problem in machine tools in general. Examples of noise control by modification to the surface structures, with due regard to the acoustic radiation efficiency, were available from the engine work and a few machine tools and presses. This was sufficient to teach the principle, but not to design optimal structures. The participants could try isolation for stiff covers and vibration damping for flexible covers. Suitable materials and suppliers were listed.

Local shielding over noise radiating surfaces is a technique which appeals to many production and works engineers; and when used insightfully it can be effective.[2] The identification and ranking of radiating surfaces (acoustic sources) was described in the diagnosis section, so apart from refreshing the memory, the main part of this section was aimed at choice of suitable materials, finding suitable mounting points on the plant, use of absorbent, and above all avoiding the construction of a bigger and better noise radiator!

Noise control of pneumatic exhausts, compressor intakes, fans, etc., was tackled together by describing the different types of "silencer" and the uses for which they were most suitable. This section is an important one for factory noise control, as unsilenced exhausts, poorly designed air jets and the air blasts used to remove product from press tools contribute a lot of noise. The participants were encouraged to tackle these first, as the amount of air used could be reduced and useful savings made.

The last section covered acoustic enclosures and the use of acoustic absorbent in enclosures. The material is well known, but the presentation was chosen to end the course in a very positive way.

5.7 Presentation

The contact time for this course was 7 hours. Each session was limited to twenty minutes maximum, followed by five to ten minutes for questions.

Two Noise Control Engineers presented the material acting as a team. In addition, recent recruits were introduced to their clients and colleagues by presenting parts of this course. All team members would take part in the question sessions and close teamwork and rapport was essential to impart the material clearly in the time available.

Spare demonstrations and some more examples were available to further emphasise key points in response to questions. The team would run these to tailor the course to participants' questions.

All the visual aid material was on 35 mm slides, in a carousel, to save time during presentation. A white board was used to illustrate answers to questions. All the audible demonstrations were recorded, in order of requirement, on one stereo tape, with voice-over introductions and colour-coded leader tapes spliced in to find the exact start of each demonstration while the lights were dimmed for the slides. An "off-duty" team member set up the tape replay system. The lights were raised for the practical demonstrations and for questions to avoid long dark periods.

Each section had between one and three demonstrations to emphasise the key point(s). In many cases, some piece of hardware would be circulated to show practical engineers some real detail. Examples were stampings from a lamination press with staggered tooling, scrap stamped out by a scalloped punch, a damped engine rocker cover, a helical gear and anti-vibration mounts.

The final session of the course was devoted to acoustic enclosures. By this time, some of the participants were beginning to feel tired. After a short introduction describing the effects of panel resonances and coincidence (very simply indeed) with a set of "sound attenuation" (SRI) data plotted to show how different constructions contained noise, a "sound-proof box" demonstration was described. This demonstration incorporated numerous ideas and at least two puzzles. One of the participants near the front was given a sound level meter and another was appointed "scribe". The attenuation of a noise source by enclosure was taught by a series of brief experiments (single leaf, double leaf, leakage, noise inside, absorbent, leakage again, anti-vibration mountings, etc. with the demonstration). During this session, the participants gradually took over, as they wanted to try various ways to improve the "box." This left the participants in a very positive mood having jointly solved the puzzles in the demonstration.

The Noise Generating Processes chart was gradually modified to a Noise Control Solutions chart to summarise the key points at the end of each section. The completed Solutions chart formed the summary for the whole second part of the course.

Shortly after issuing the noise requirement for new machines, we ran a Noise Control Course for suppliers of plant and equipment. We anticipated that the participants would come with their critical faculties well-honed and we were not disappointed. This was one of the most interesting courses we have ever run. The discussion periods were packed with questions, experience, counter-experience and several challenges. Most of the delegates were experts in their fields and we learned a lot about machine tool design that day. The work invested in the course design and numerous examples proved to be necessary and just adequate. At the end of the day, several of the 100+ delegates expressed their appreciation for the initiative, material and presentation. Some were kind enough to say that the day had caused them to think afresh about the designs for which they were responsible.

5.8 Effectiveness

As before, there was no formal examination of the knowledge which the participants had acquired; and the real test of effectiveness was the reduction of noise in Company premises. The awareness of noise control treatments allowed Methods engineers and Production engineers to press for cost effective noise control when ordering new plant and equipment. The same treatments were installed in similar machine tools during refurbishment. No machine tools and fewer presses were incarcerated in sound-attenuating enclosures. The noise in

Company premises fell steadily until hearing protection was needed in very few places (mostly unattended spaces such as compressor houses and special purpose plant). The engineers on each site applied and maintained noise control measures such as air exhaust silencers, vibration isolation for small items of plant, replacement of noisy power transmissions with quieter drives and some innovative process changes. After seven years, an independent survey of factory noise in one of the divisions the showed that almost all the major noise problems had been tackled successfully. The machine tool suppliers collaborated fully with the Company in noise control at its origins.

6 CASE STUDY 3 MSc MODULE: NOISE AND VIBRATION CONTROL I

6.1 The Students

Most of the MSc students approached the subject with no experience of noise control problems and, apart from most of the students taking the automotive courses, many intended to pursue a career in research. Most of the students came from outside the UK and for some, understanding English was hard work. For these students the mathematics was much easier to understand.

The author had the opportunity to try an automotive version of the industrial Noise Control Course early during his period at the ISVR. Even for a class of students who were mostly from the UK and who had, perforce, more extramural practical automotive experience than they felt they needed, the course was hard work. They had to assimilate a design outline of each component (engine, gearbox, alternator, hydraulic pump, etc.) before they could study the noise generating mechanisms and potential noise control options at the origins of the noise.

Students from conventional courses required a full description of any machine that was to be tackled even when related to the noise problems they heard around them. They felt comfortable with an academic course structure and some had selected the course for its academic content.

6.2 Course Design

The study of Acoustics and Vibration had formed the basis of courses leading to a Masters degree for over 25 years in the Institute and many variations had been tried. There was a belief that the students would benefit from lectures from someone with a background in practical noise control and the academic staff were most helpful in providing material and the rationale behind the existing course design. The first semester module was one of several introductory modules and had to contain enough preparatory material to enable students to choose freely from the courses in the second semester. The syllabus could be extended but not reduced.

An opportunity occurred to discuss modelling, in sound propagation through porous media (acoustic absorbent). Several simple models exist, which are based on alternative bold assumptions, and which can be used in appropriate applications. These can be used also to generate a useful debate on modelling, however this point is reached shortly after the students have become acquainted with the teacher and it needs very careful handling. Students who start from the premise that everything they are taught is somehow going to hold true in all circumstances; can find such a debate confusing as their precepts are challenged. The aim is to promote the use of models in the full realisation of the implications of the assumptions in their construction.

A progression from very simple models to more complicated ones was suggested where relevant, as preparation for the more complex situations to be tackled in the next semester. Discussion of the choices to be made when doing "back of an envelope" calculations was encouraged to promote a more critical view of the course material; since making such choices is a start towards preparing students for real problem solving.

The importance of selecting an appropriate model from the literature, or their own research, takes time to impart; more time perhaps than a prescriptive instruction. However, the benefit to the students is that industry perceives them to be more mature technically. Mature assessments are needed to contribute a reasoned argument for a course of action, to gain the requisite backing and support from senior managers in industry, to provide useful predictions in practical situations and to develop the predictive process in the light of conflicting experimental results and experience.

6.3   Presentation
Wherever relevant, practical demonstrations were used to illustrate and reinforce key points. To take a simple example, standing waves can be illustrated with a sine wave generator, amplifier and loudspeaker. Suitable standing waves can be found in each of three dimensions near 1 kHz where the changes for small movements of the head are noticeable. This provided also 3 minutes of relaxation in the middle of a lecture period to improve concentration.

When acoustic absorbent was discussed, some pieces of suitable fibrous and open-cell plastic foam were handed round. When Vibration isolation was discussed, a selection of isolators was shown with comments upon their applications and design.

6.4   Assessment
The assessments were by conventional examination papers

6.5   Effectiveness
The automotive students, who took this course, took a second semester course which used the linear acoustic models in this course as a starting point for muffler design and another which discussed non-linear hydraulic wave action in fuel injection. These students seemed to find the 2nd semester course easy, however the effectiveness of this module is hard to assess in isolation.

7   CASE STUDY 4 MSc MODULE: NOISE AND VIBRATION CONTROL II

In the second semester, students selected modules which included Acoustics II and Vibration II as well as N&V Control II, so the students had more commitment to noise control as a career option.

7.1   Course Design
The course design was less constrained by the need to integrate with other modules. The course started with noise control at its origins, with an analysis of how each treatment can be

designed to suit particular applications. The examples from the Noise Control Course were very useful to indicate the range of application. With the benefit of hindsight, it could start equally well, if not better, with a more detailed discussion of the conventional noise and vibration control options: enclosure, duct silencers, vibration isolation, etc., and work backwards towards the origins of noise and vibration. The central theme was that there are at least three options to control noise: at its origins, in transmission and by enclosure and that the most cost-effective course of action should be chosen after a systematic appraisal.

7.2 Presentation
Tape recordings, associated slides and actual components were used to illustrate each concept.

7.3 Assessments
Examination by conventional questions in a set paper is a part of the infrastructure of educational establishments, providing a time structure and motivation for students and staff alike. However, if the questions require the sort of analysis needed to solve practical problems, where alternative options are available, the examination time constraints prevent a meaningful discussion of different options. Furthermore, if the question is open-ended, the student may fail to find a good solution and not be able to demonstrate their learning achievements adequately. Hence, even if the questions are inspired by real problems, fitting an answer into a proportion of the examination period and avoiding traps for the unwary is very difficult.

After considerable discussion with colleagues and long discussions with the external examiner, an alternative approach was tried. Three noise control tasks were identified as suitable to examine the students. The requirement was discussed with the students. They would analyse three noise and vibration problems and devise a practical solution for each. Their analyses, discussions on alternatives and recommendations were to be written up in the form of a report with all the information necessary to support their recommendations. The analyses required to fulfil the academic requirements were written as appendices to the main report. The report itself was be in two parts; the first being an executive summary on a single page which was addressed to someone who has to sanction the expenditure. The second part contained the background, justification and detailed recommendations with engineering sketches of their solutions, which was to be addressed to their peer group. The students were shown three industrial reports describing successful noise control investigations which embodied all these features.

The first assessment required the students to recommend noise control measure for a 7-hp direct current traction motor. They were given a sketch of the cross section and to add further realism, left in charge of a complete motor and a spare armature. The task was to reduce the noise by 3 dB(A) initially, but a second customer might purchase more motors if it were made 6 dB(A) quieter. The reports the students produced were of commendable quality and the recommendations were good; however, they concentrated on one line of attack.

Two experienced engineers had tackled the same task previously in the Noise Centre; and as a result a prototype motor existed which was 9 dB(A) quieter. After the students' reports had been marked and the marks returned to the Examinations Office, the project was discussed with the students as a group. The prototype was shown to the students for their comment.

The second project was to specify the anti-vibration mountings for a pump test rig (Fig. 5). The mountings were about to be ordered for this rig, so the project was "real". The rig incorporated vibration-absorbing structures, which was germane to their studies. The second moments had to be calculated from measurements of the rig in situ and coupled modes were entirely possible. The students chose the mountings which were ordered.

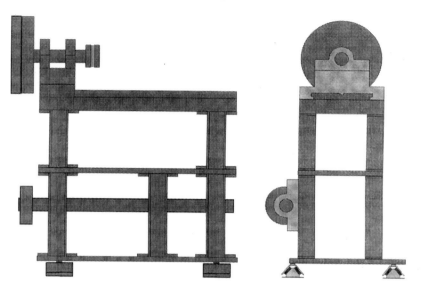

Figure 5 Quiet Drive Rig which required anti-vibration mountings

The third assessment was a challenge to solve a widespread problem in offices; the partitioned office which is anything but sound-proof. By good fortune, the offices in the department illustrated all the routes by which noise travels around partitions. The sketch supplied to the students showed a cross section of the office (Fig. 6) and the accompanying sheet specified the partition construction, air inlet and outlet ducting, heating, doors, etc. The task was to improve the sound attenuation so that the secretary outside would not be able to understand conversations, even when they were made via a loudspeaker in the telephone apparatus. The requirement was specified so that the students had to do some research to find the acoustic specification for the office partitions and the flanking paths.

As before, the reports were all of a high standard, although the costings were rather thin. The analysis of each problem was thorough and perceptive. The solutions were competent and began to show an understanding that practical details had to be specified. The students were particularly keen to reclaim their work after the external examiner had seen it.

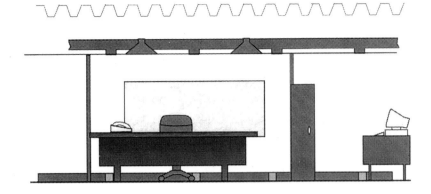

Figure 6 Office noise insulation, to ensure privacy for conversations via a loudspeaker telephone.

Each report took around one hour to read, assess and comment upon, so the time investment is six to nine times that for marking a conventional examination paper. However, the students seemed to value the comments on their work and went of with some "templates" for future tasks.

8   SUMMARY AND COMMENT

By comparison with other professions, engineers are employed in a wide variety of roles in many functions in industry; and their career patterns are just as diverse. As an example the roles of engineers responding to varying customer demands for engineering expertise in one large and innovative engineering company are summarised in Figure 1. This chart shows also how the specialist noise, vibration and dynamic analysis group interacts with the multiplicity of processes which are running simultaneously in other engineering departments. This presents quite a challenge to those responsible for providing appropriate professional education and training.

The process for developing new products to meet changing market requirements and legislation in one company, as summarised in Figure 2, is changing as computer-aided engineering advances.

From discussions with others in the automotive industry, most graduates need further intensive training upon entering their first job. (This was the function of graduate and student training programmes.) This places a 6 to 9-month training burden on their first employer.

Employers expect their engineers to move between the various functions in their company, as the need arises and to further train engineers for senior positions. Graduates who learn rapidly are best fitted for this career pattern. Senior engineers and supervisors expect a rigorous education in the fundamentals of engineering, as preparation for such a career.

Further specialist education may be required for a career in noise and vibration control. Only students who decide early that they will spend a large part of their career in practising noise or vibration are likely to invest in a specialist first degree or Masters degree. MSc degree courses offer a route for graduates to specialise after some industrial experience, which is more attractive to employers. However, the employer needs to provide a powerful incentive for the graduate to return afterwards; and promises made in good faith may be difficult to honour a year later.

For industrial training, short courses are really successful when they just precede the task. The educational status of the participants, including the formal education, can be found from their job specifications to direct the level of the material and style of presentation with some precision.

An intensive course needs a carefully constructed introduction, directing the participants towards a profitable use of their time and providing information on the level and pace of the course. The introduction to the in-house noise courses contained a form of "contract", in that the specified material will be presented with a well-defined structure and related to real problem solving. The deliverables were related to this "contract" as the course progressed.

The introductions to the noise courses were designed to motivate the participants to learn. By contrast, students do not seem to be aware of the structures of the academic courses they take and their motivation to learn is not assisted by any clear perception of the value to a future employer.

It seems to the author that much of the debate on relevance of education to the needs of industry centres on the material in academic courses and not the presentation. With existing material, the presentation can be adapted to incorporate, for example: -

1. A realistic view of modelling to promote its wider use.
2. The importance of validating models with suitable experimental data.
3. Compact case studies with real hardware wherever practicable.
4. Practice in writing reports in a form which is more useful in industry and with the necessary academic rigour.
5. Challenges and some intellectual reward in solving relevant problems.

In addition, apart from a thorough grounding in the fundamentals, graduates seems to need some education and training in survival skills, for example: -

1. A reliable process for tackling tasks and problems successfully.
2. Active listening and examining.
3. A learning process with which they can pick up and master a new subject quickly.
4. An ability to work effectively in a team.
5. Practice in reducing data, putting results into context and interpreting them.
6. Practice in communicating arguments and recommendations concisely and attractively.

Thus equipped, engineering graduates can adapt themselves to tackle unfamiliar tasks with confidence and ultimately job satisfaction and fulfilment.

9 ACKNOWLEDGEMENTS

The author thanks the directors of Lucas Diesel Systems for permission to publish this paper. He is grateful for all the help, constructive criticism and advice he received from present and former colleagues including Chris Best (for Figure 1), Steve May, Carol Shaw, Peter Wilson, Steve Worley, Ron and many others in Lucas Industries Noise Centre.

The author is most grateful also for the support of IVECO and the Royal Academy of Engineering for the European Chair in Automotive Engineering in the Institute of Sound and Vibration Research at the University of Southampton. The author would like to take this opportunity to express his thanks to his ex colleagues and students in the ISVR who were generous with their advice and help including Dr David Anderton, Dr Nick Lalor, Professor Frank Fahy, Professor Chris Morfey, Professor Jo Hammond, Professor Phil Nelson, Dr Mike Brennan, Dr Roger Pinnington, Dr Neil Fergusson, John Dixon and the engineers and technicians of the ADAU, the staff of the ICS, and others too numerous to mention.

10 REFERENCES

1) Russell, M F "A comprehensive programme for hearing conservation and noise control" SAE paper 870954 proc. Soc. Auto. Engrs. Noise and Vibration conference Traverse City Michigan April 1987.

2) Russell M F "Reduction of noise emissions from diesel engine surfaces" SAE paper 720135 Auto Eng Congress Detroit Jan 1972

3) Russell M F and Cavanagh E C "Establishing a target for control of diesel combustion noise" SAE paper 790271 in P-80 Proc. Diesel Engine Noise Conference SAE Auto. Eng. Congress Detroit Feb 1979 pp 89 - 101.

4) Russell M F "Automotive diesel engine noise and its control" SAE paper 730243 SAE Int. Auto. Eng. Congress Detroit Jan 1973 and SAE Transactions Vol. 82 1973

5) Wilson, P M "A pragmatic look at sound in real factory spaces" Proc. Inst. of Acoustics conference Noise Control in Factory Buildings June 1982.

C574/028/99

Sustainable learning in the automotive supply chain

N BARLOW and A C LYONS
Liverpool John Moores University, UK
P F CHATTERTON
Daedalus, London, UK
A GLOVER
The Technical College Birmingham, Tyseley, UK
M JONES
University of Hertfordshire, Hatfield, UK
B OXTOBY
SMMT Industry Forum, Birmingham, UK

ABSTRACT

For a number of years now a group of representatives from the Higher and Further Education sectors in the UK have been working in collaboration with the SMMT Industry Forum and the DTI Automotive Directorate together with Daedalus, to establish programmes of sustainable learning in the automotive supply chain. The learning is focused on the SMMT Industry Forum's common approach toolkit for achieving sustainable quality improvement in the UK automotive supply chain. The success of the learning is measured against the industry's seven key measures which are described in terms of quality, cost, delivery and partnership. Two pilot schemes have been undertaken with a number of automotive supplier companies based in the Midlands. The initiative is supported by an on-line Information Management System that is used to enhance collaboration between the parties. It is accessed using a Web browser on the Internet (in a closed user group) and allows best practice and resources to be shared as well as promoting efficient and cost-effective management of the initiative. The developments up to now have concentrated on the tools of project management and variability reduction through statistical process control. There are plans to expand this activity on a regional basis during 1999, with involvement from other partners, and also to enhance the Information Management system with on-line mentoring support and resources for the learning groups.

1. INTRODUCTION

The automotive supplier companies recognise that if they are to remain competitive in the global market, which has been established by the vehicle manufacturers, they have to invest in a sustainable competitiveness process. The SMMT Industry Forum have identified a set of QCD (Quality, Cost and Delivery) measures against which supplier companies can benchmark their performance. These measures are identified as:

> Non Right First Time
> People Productivity
> Stock Turns
> Delivery Schedule Achievement
> Overall Equipment Effectiveness
> Value Added Per Person
> Floor Space Utilisation

In order to bring about improvements in these measures the Industry Forum have identified a set of tools and techniques. These include:

> demand levelling
> flowcharting
> waste analysis
> standardisation
> visual control

Through their Process Improvement Master Class, Industry Forum have brought about significant improvements in the supply chain. It is recognised that the major competitive edge a company has is within the knowledge and skills base of its workforce. To this end, with support from the DTI Automotive Directorate, Industry Forum have brought together a group of universities and further education colleges to develop a sustainable learning process based on their tools and techniques. Over the last eighteen months three academic partners, Liverpool John Moores University, University of Hertfordshire and The Technical College Birmingham have developed and piloted the delivery of two of these tools/techniques with some thirteen supplier companies based in the Midlands. The two tools/techniques chosen for the pilot schemes were Statistical Process Control (SPC) and Project Management.

2. A SUSTAINABLE COMPETITIVENESS PROCESS

It was recognised at the outset that the transfer of knowledge would not be sustained within the company unless there was a commitment from the top which would also involve the introduction of company standards and only if the process of knowledge transfer was deemed to be a success.

Throughout the project a framework was used to plan and review the implementation of the programmes of learning in pursuit of improvements in the quality and delivery of products and processes and also reductions in cost. The framework or process was developed after studying best practices in automotive component suppliers by talking to and working with the leaders in the recipient companies. The framework, referred to as a Sustainable Competitiveness Process, is shown in Figure 1.

Figure 1 Sustainable competitiveness process

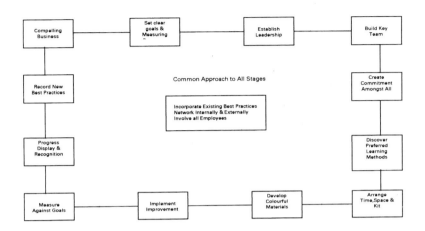

This process was used to:

- identify the strengths and experience of the subject that already existed in the company and also to specify the measurable goals to be targeted.
- diagnose existing business performance and the potential of the infrastructure to be used or needed for transferring the learning from the training to work in progress.
- pinpoint the leadership needed to sustain momentum and apply the tools and techniques for developing a factory environment which was conducive to change and growth.
- brief the delivery teams upon local circumstances which would allow for the design of the knowledge transfer to be tailored to suit the company.
- assess the effect of unforeseen occurrences upon the total activity and to take necessary action and also to capture learning points for spreading best practice.
- to project manage the total programme and remain as a resource to secure sustainability in the longer term.

3. PILOT PROGRAMMES IN PROJECT MANAGEMENT

A set of four Project Management programmes ranging from Business Strategy to Starter Practitioner were designed. Individual topics overlapped and programmes were offered with the view that detail would be adjusted when the company needs became known. All programmes except one were 2 days duration with an option for half day sessions. A two week gap between delivery days was recommended for in company work to be undertaken.

Two companies, a general component fabricator and a specialised "process" component manufacturer were selected for the first pilot. One was known from previous delivery of a

Project Management foundation programme by one of the academic partners. Their Chief Executive and Finance Director requested a high level course for business planning in half day deliveries. The second company opted for the mid level programme. Its highly qualified delegates and business managers were interviewed during a full day visit by the delivery team. Each delivery team consisted of a university lecturer and a further education lecturer. Sessions were prepared and delivered jointly according to tutor strengths and preferences. Between twelve and fifteen delegates with a good balance of seniority attended each course and both programmes were highly interactive. The companies declared valuable useful outcomes but the delivery team identified the following problems:

- Earlier Project Management work at the first company had not been sustained. Staff were not planning and working in project teams and were not able to take full benefit of the high level programme. The programme plan was changed at the mid point to emphasise project planning and teamwork basics.

- The second company was hard driven by all vehicle manufacturers' projects and needed to bring in key process updates. They needed to develop a process to classify and manage all projects and to cover strategy and resourcing not just the time and cost needs of the vehicle manufacturers. Delivery was adjusted to cover their need.

- The different style of delivery between tutors became apparent in these programmes. Best practice requires the "knowledge and material" emphasis of the university to combine with the more informal "class learning" of a further education college. Team teaching makes it essential that tutors can adjust to these differences, and two tutors working interactively is the preferred approach.

- Diagnostics with top management alone are less likely to identify essential needs. Even with a diagnostic day, flexibility has to be maintained during delivery with a high degree of delivery partner trust

The second company has proved highly proactive in forming and operating its own Project Management centre from programme guidelines. Recent presentations suggest that their methodology now controls a wide range of projects overriding differences in individual OEM's. The first company is still dominated by OEM drivers and reactive to these. Following the pilot programmes it was agreed that a single programme framework would be produced covering all aspects of project management from classification to detailed planning. Delivery teams would then provide detail around this framework based on diagnostics data.

A second pilot programme was delivered with support from European funding to a further six supplier companies each with some 150 to 250 employees. Three of these companies were general purpose fabricators, two specialist die casters, and one a plating/heat treatment process company.

All of the programmes were delivered in company and the presence of company directors helped to minimise disturbance. The programmes were highly interactive and challenging with high flexibility demanded for company agendas and examples. This was possible with the common framework within which material could be added or subtracted. Diagnostics days helped team credibility and established company products and processes with Project

Management experience. High level participation reinforced messages and problem areas. A key outcome of the programmes was the ability to distinguish between "reactive" customer generated projects and infrastructure/business change projects. In all cases companies were finding it difficult to sustain or even to generate medium to long term projects critical to their survival and competitiveness because of the short term emphasis on customer needs.

Work with these companies enabled typical customer order projects to be analysed for efficiency and control in line with Industry Forum drivers. Probably the most valuable outcomes however lay in the generation of a more proactive forward looking approach to company infrastructure and business projects. Many companies have indicated their need and willingness to be proactive through Project Management in order to override what they frequently observe to be inefficient customer driven processes.

4. PILOT PROGRAMMES IN STATISTICAL PROCESS CONTROL

Effective statistical control of processes has been identified as critical to automotive companies achieving world-competitive levels of business performance. This programme was specifically designed to embody those statistical process control (SPC) and process improvement techniques that if judiciously and competently applied, and approached in a manner that is structured and conducive to user involvement, have consistently proven to improve quality and productivity within the automotive components sector.

The approach was geared to actively engaging top management in the sponsorship and direction of the programme and user know-how and involvement in its implementation. This *top-down commitment, bottom-up implementation* embodiment of the approach has proven to be the most vital ingredient to success. The commitment of the highest-level decision makers is essential for the wellbeing of the initiative. Only those directly involved in establishing business direction can ensure that adequate time and resources are committed to the programme and bring about the necessary conditions in order to sustain the momentum the programme creates. An overview of the key process elements is shown in Figure 2.

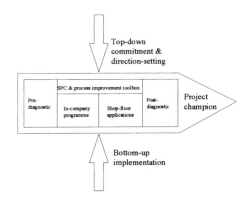

Figure 2 Statistical process control approach

In addition to top management involvement, the approach to the SPC programme made use of a project champion to facilitate the training and ensure the sustainability and continued success of the application of SPC and process improvement tools and techniques within the company.

The specific content of the programme was tailored to the needs of the host company. The target audience ranged from SPC novices to architects and exponents of mature SPC programmes. The inclusion and articulation of users' (participants') requirements in planning the programme content ensured that the needs of those affected by the process were met, and participants are equipped with the necessary tools and techniques to be autonomous and effective in controlling process variation within their own environments (*bottom-up implementation*). With such involvement, users and managers are much more likely to accept changes to current working practice. It has been much easier for the participants to accept new ideas and changes having had a significant involvement in their development.

An indication of a typical delivery schedule for the programme was as follows:

Session 1	-	the business case for SPC; process definition; process mapping; existing in-company controls;
Session 2	-	problem prevention: failure modes and effects analysis; variation; fundamental statistical concepts; common and special causes of variation;
Session 3	-	introduction to control charts; levels of control; measures and parameters; critical and significant process parameters and product characteristics;
Session 4	-	process capability; capability indices; capability studies;
Session 5	-	types of control chart; attribute and variables data;
Session 6	-	SPC implementation and use; process improvement;
Sessions 7-12	-	case studies were undertaken by the programme participants in their respective companies under the guidance of the programme team.

The programme is typically delivered over 8 to 12 successive weeks consisting of three hours expert tuition, practical demonstration or in-company application. At least the last third of the delivery period was devoted entirely to process improvement applications within and of direct interest to the host company. The mode of delivery was flexible to meet company needs but programmes were usually delivered on company premises. The delivery was supported by the on-line information management system and a suite of process improvement and SPC learning resources.

World-competitive sustainability is a major aim for the UK automotive supply base. This programme is providing participating companies with a crucial step in the right direction to achieve that ambition. Specific and quantitative benefits from the pilot programmes have so far included a 30% reduction in process scrap and a 25% reduction in lost production time

5. ON-LINE INFORMATION MANAGEMENT SYSTEM

An On-line Information Management system was developed to

- support the administration and management of the project
- support the development and maintenance of the programmes
- support the knowledge transfer to the companies

A survey of small to medium enterprises (SME) had shown that Internet access was not widespread. The decision was taken to initially adopt information and communication technologies (ICT) applications to support the administration and management of the project and the delivered programmes. It is the intended aim, as Internet access becomes more widespread within companies, to extend the use of the ICT applications in the support of the knowledge transfer process.

5.1   ICT to support the administration and management of the project

In practical terms, the project team was not able to meet face-to-face on a frequent basis. A strategy was adopted to utilise the World Wide Web (WWW) as the primary means of communications and information management. To this end, an Online Information Management (OIM) system was developed to support the project.

The OIM system is accessed using a standard Web browser and requires users to supply a User Name and Password in order to gain entry. It is divided into a number of areas, the most heavily used being the "**Discussions**" area. This works rather like an Internet Newsgroup or bulletin board where users can post discussion topics and reply to others' discussion topics. The openness of the "discussions" to the whole group encourages the sharing of information and ideas. Topics of discussion are highly varied and range from the development of the project to the need for, and techniques for, encouraging innovation within the companies.

A number of other areas have been developed within the OIM system. The aim of these areas is to support the processes of a multi-partner group that involve electronic communications and information flows. For instance the "**Meetings**" area allows any of the users who are arranging a meeting to post details of the agenda and participants. Once the meeting is complete, the minutes would then be posted alongside the meeting agenda which could be typed direct onto the system or "attached" as a word-processed file. In this way, a permanent record is maintained of all meeting agendas and minutes.

Another key area is the "**Tasks**" area. The project team has recently appointed a Project Manager and this area aims to help him to assign and track the progress of tasks. When any user views this area, they can see all of the tasks which have been assigned to them, together with the completion date, flagging up tasks which are overdue. Users can "respond" to tasks (rather like in the "Discussions" area), for instance, by reporting on progress or making a query. Like the "Meetings" area, a permanent record of all tasks and their progress is maintained and the Project Manager can easily ascertain any problems as they arise.

The "**Documents**" area is used to store all documents that are produced by the project team, such as project plans, proposals and conference papers. The "**Library**" area contains many useful references and links to sources of information such as books, journals, news articles and Web sites. It is hoped to build this area into a significant resource or knowledge base for the automotive community leading to benefits to the companies, education establishments and government and industry bodies. All users of the OIM system are encouraged to make contributions to the "Library" area.

Other areas of the OIM system include an area for maintaining records of the companies and the beneficiaries (trainees) and an area that contains a number of feedback forms which allow users to provide feedback via structured questionnaires.

The OIM system has now been in operation for six months and is actively used. One of the major advantages cited by one user is the ability to log on at any time and anywhere in the world. It has certainly achieved its aim of cost-effectively improving communications and information flows within the project team.

5.2 ICT to support the development and maintenance of programmes

The project team have commenced a project to develop a range of high quality learning materials as a collaborative venture. The materials will be in the form of a tutor toolkit where tutors can "pick and mix" components of the materials to suit the customised nature of the programmes. Materials will be in electronic form, giving the tutor the choice of which media to use to deliver the materials, e.g. hard-copy, Web page, Adobe Acrobat file, overhead PowerPoint presentation.

The OIM system has been extended to aid the process of developing these materials by allowing developers to post draft versions on to the system and where reviewers can post responses, comments and requests for revisions or "sign-off" the draft as the final version. The OIM system automatically version-controls the drafts and provides an audit trail of the development cycle. The same techniques will be used to maintain new updates of the materials.

5.3 ICT to support the delivery of programmes to the companies.

It is intended to provide an online "Support Forum" for the companies where tutors and trainees can collaborate together in a similar manner to the "Discussions" area of the OIM system. This will allow tutors to provide online "mentoring" support after the cessation of the formal part of the training programmes and will also allow members of different companies to collaborate together. This phase of development will require keen attention to its introduction with comprehensive training and ICT support. It is also hoped to develop an online "Project Management Role-play Simulation" which will allow groups of trainees to "practise" project management scenarios.

A by-product of the use of Internet applications in the learning programmes will be to teach the companies about innovative Internet-based techniques, thus helping to support the Government's ICT aims within the competitiveness White Paper.

The long-term vision of the ICT applications is to develop an online automotive community where supplier companies, vehicle manufacturers, the SMMT Industry Forum, Universities and Colleges and government are actively collaborating together, exchanging and sharing knowledge and working and learning together.

6. CONCLUSIONS AND FUTURE DEVELOPMENTS

The realisation that gaining a competitive edge for any business today is directly related to the involvement of the people within that organisation, has brought new challenges to colleges and universities involved in supporting the automotive supply chain. The present partnership involving Further Education, Higher Education, Industry and Government has been working together for a number of years developing a successful delivery programme supporting sustainable learning in companies.

The present partnership has a proven record of success with training employees in Statistical Process Control and Project Management and incorporating these tools into company practices, ensuring sustainable and quantifiable improvements within the organisation. These improvements have been quantified in terms of Quality, Cost and Delivery of company products and services. The partnership between Further and Higher Education has developed a common approach to the delivery of these two tools. At present a common language is being developed and will be published to ensure continuity of the delivery content and process as it expands to other regions and incorporates other institutes. The aim of this published material is to provide a framework for tailor made courses to meet individual company needs for improved manufacturing performance, product development and company development.

The development of these two tools have been mainly funded by the European Social Fund. It is hoped that other training modules will be evolve as demanded by industry to form a comprehensive toolkit. It is our objective to continually develop the range of products and services available within this toolkit.

To date, the two training tools have been delivered to the automotive sector in the West Midlands. It is envisaged that through the help of continued European funding delivery can be expanded to the North West and Eastern regions where the other educational partners are based. In coping with this expansion, it is planned to widen the partnership to include other like minded, high quality, committed institutions. This will involve each existing educational partner linking with a local complementary establishment strengthening the unique HE/FE partnership. Links to other government organisation and industry representatives in those regions are also being forged. It is our goal to have a comprehensive network of partners covering all of the manufacturing centres in the United Kingdom. All of the partner institutions would have rigorous quality assurance procedures in place, assuring the quality of the delivery and those staff engaged in the process.

In using the latest Information and Communications Technologies (ICT), has enabled the efficient managing of this project, especially since partners are dispersed throughout the country. The development of Information and Communications Technologies (ICT) to support sustainable learning will be a key part of any future developments. The plan is to ensure that ICT is an integral part of the learning experience offered to participants and companies, and actively supports the management of the learning process.

The partnership is currently investigating suitable accreditation for existing and proposed training modules within the toolkit that is hoped to be available in the future. These qualifications will not only recognise individual students achievement, but it is hoped that they can be linked to national standards. The content of the toolkit will be developed to include industries requirements were possible and will be linked to existing courses within participating institutes. This will help to address the issue of education relating to industry requirements and encourage students to engage in life long learning.

The present partnership has a proven track record and it is believed that this can be successfully expanded and shared for the benefit of industry in this country, which is in need of supported sustainable continuous improvement to secure its existence.

8. ACKNOWLEDGEMENTS

The project goals would not have been achieved without the support and help of Francis Evans, Keith Jordan and Nigel Goulty at the DTI Automotive Directorate and also Graham Broome, the Chief Executive of SMMT Industry Forum. We acknowledge also the contributions of Professor Peter Bullen and Professor George Harland from the University of Hertfordshire and Rita Davey from the Technical College Birmingham. This project has been funded in part by the European Social Fund.

Undergraduate

C574/031/99

A design-and-build racing car project: the changing face of automotive engineering at the University of Leeds

A J DEAKIN, P C BROOKS, M PRIEST, D C BARTON, and D A CROLLA
School of Mechanical Engineering, University of Leeds, UK

Synopsis

In recent years, students at the University of Leeds have been involved in the design and build of a Formula-style single-seater racing car. The students concerned have been exposed to the latest analytical techniques and technological innovations in motor sport engineering. Moreover they have gained valuable first-hand experience of effective team working, internal and external communications, and completing a realistic engineering project to tight financial and time constraints. Complementary to the race car effort, new MEng and MSc programmes in automotive engineering have been developed to meet the changing expectations of industry and the professional institutions. This paper discusses the impact of these developments on teaching and learning outcomes for automotive engineering students at Leeds.

1. INTRODUCTION

The School of Mechanical Engineering has three years of experience in running a major student project to design and build a racing car to compete in the well-known Formula SAE competition in the USA The project has proved to be an ambitious undertaking! Nevertheless, it has also proved to be a powerful motivating factor in the students' learning experience in professional engineering. The growing success of this competition and its spread, firstly to the UK as the Formula Student event and increasingly on a global scale, are evidence of this motivational aspect. The paper aims to analyse, specifically, the impact of this project on the overall learning experience of the 120 undergraduate and postgraduate Leeds students who have participated so far, as well as the associated benefits to the School.

In parallel with the race car project, the School has recently introduced an MEng programme in Automotive Engineering which aims to meet the requirements of the professional engineering institutions as exemplified by SARTOR 3, as well as providing an attractive alternative to mainstream mechanical engineering for students wishing to pursue a career in the industry. A taught postgraduate MSc programme, with the same title, has also been developed which provides a conversion route for students with a first degree in a related discipline as well as more specialised training for potential researchers in automotive related topics. The high quality teaching materials developed for this programme provide a basis for short courses and distance learning opportunities in line with the changing model of EPSRC funding of advanced taught courses. The paper discusses these developments and the links with the Formula race car which acts as an important focus for project work associated with both MEng and MSc programmes and offers the closest experience to real world engineering responsibility that can be provided within the confines of a higher education institution.

2. THE RACING CAR PROJECT

2.1 Overview of Formula Student Race Car Competition

Formula Student is a design-and-build competition where teams of students are challenged to produce a small, high performance, single seater racing car. Formula Student is the European sister event to the well established American Formula SAE competition which currently attracts 100+ universities from around the world. The solution which encompasses the best compromise between performance and cost whilst bearing in mind a required theoretical production run of 1000 vehicles per year will win; this is not necessarily the fastest car. All designs are judged by a team of professional motorsport and automotive engineers. Teams must produce both a detailed cost report on the manufacture of the vehicle and a presentation selling the vehicle to a potential manufacturer. Following the static appraisal of the vehicle, the performance in terms of straight line acceleration, cornering ability and time against the clock are assessed. The competition culminates in an exciting endurance event where the cars race for half an hour to prove their reliability and performance. Figure 1 shows the current Leeds car in action.

Figure 1: The 1999 Leeds car on the track at Formula Student '99

2.2 Historical Development of Formula Car Project at Leeds

Year 1: A team of 5 MSc students was formed to investigate the process of designing a racing car from scratch, including defining the overall requirements and laying out the engineering concepts. The required engineering analysis techniques were identified including finite element analysis, vehicle dynamics, kinematics and other CAE tools. By the end of this year, a baseline design was established from which the first UK car could be built.

Year 2: A team of 18 undergraduates was assembled to execute the daunting task of detail design, manufacture and test of not only the race car itself but also an engine test-bed. To the students, this meant they had to master, firstly, new engineering design techniques, secondly they had to incorporate their findings in a detailed design which, thirdly, they had to produce themselves; finally they had to test and prove that their designs worked. The resulting Mark I car consisted of a simple space-frame chassis with bonded-in, structural panels and a rough and ready carburetted fuel system. Although the car was overweight and unrefined, it proved a good solid basis from which to work and was reliable enough to compete in every event in the 1997 SAE competition.

Year 3: Here the goals were focused to allow major improvements to the design using more advanced analysis techniques and a team totalling 35 students. A monocoque chassis was designed and constructed from structural honeycomb sandwich panels and a fuel injection system was developed, the whole vehicle being much more refined and having better system integration. The visit to the 1998 SAE competition proved extremely successful with the team winning the Altair prize for the 'Best Analytical Approach to Engineering Design' on account of the extensive computational modelling and validation that had been performed. In addition, sixth place was achieved in the Design category out of over 100 entrants with similarly good finishes in several other events.

Year 4: The team again grew, this time to 60 students, including first and second year undergraduates for the first time to give them early exposure to the demands of the Formula Car project. The design was similar in concept to the previous car although more sophisticated engineering analysis was carried out such as the crash simulations described below. The chassis was both significantly lightened and better integrated to produce a more competitive vehicle. The result was that the team won outright the Design event in the 1999 SAE competition. The students were also awarded first place in the safety/crashworthiness section as well as retaining the Altair prize.

2.3 An Example of Technical Development – Chassis FEA

A prime example of how students' learning experience has been continually pushed forward by the project is exemplified by the work done on chassis analysis which has been primarily conducted through the application of Finite Element Analysis (FEA) techniques. The FEA model of the Mark I car was based upon a simple beam element idealisation of the spaceframe chassis. Although this model was perfectly satisfactory for estimating the effect of changes to the design on the torsional stiffness of the chassis, it would not have been suitable for investigating the crash safety (which is considered an important element of the overall design) since experience on a similar spaceframe chassis has shown that a more detailed shell element model of the vehicle front end is required for accurate crash simulations [1]. The FEA model of the Mark II car used shell elements to represent the honeycomb sandwich construction in isolation from the surrounding spaceframe. A major issue here was how to represent the properties of the sandwich panels in a thin shell idealisation in which only the thickness and average in-plane properties of the panels can be specified. This is particularly important in a crash simulation since the complex failure modes of the honeycomb panels make simple shell element idealisations problematic. For the Mark III car, both the spaceframe and the composite sections of the monocoque have been modelled together using a combination of beam and shell elements, Figure 2. This model has been used to perform both side and frontal impact crash simulations, Figure 3, as well as torsional stiffness predictions. Separate analyses were carried out on the impact properties of the energy absorbing nosecone fitted to the Mark III car. These predictions were subsequently validated through physical drop-weight tests on an actual nosecone.

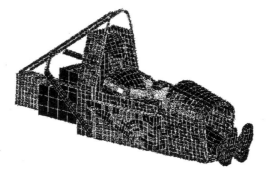

Figure 2: Finite element model of Mark III race car

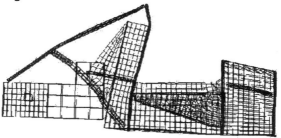

Figure 3: Deformed shape of Mark III car after frontal impact simulation

2.4 Finance and Marketing

To run a Formula Car project requires tens of thousands of pounds each year. The cost to build a car excluding academic and technical staff time and attend both the Formula Student and Formula SAE competitions is of the order of £20-25k, this being split approximately 50/50 between building and competing. Obviously such a project cannot be financed solely by the academic institution and therefore industrial sponsorship has to be sought to cover a large proportion of the cost. This task is given to the project students who have to produce the required publicity material, keep their website updated and approach industry directly to gain substantial backing. This challenges students to sell both themselves and the project to the outside world and teaches them more about the constraints of working in a commercial environment.

Even though many thousands of pounds are raised from outside, there is still a large cost to be borne by the academic institution and this cost must be justified. The activities of the Formula Car team not only act to market the project but also the School of Mechanical Engineering itself. In the last academic year, over one thousand brochures have been distributed to industry explaining not just the project, but some of the strengths of the School. In many cases, this has led to new industrial contacts as well as a strengthening of existing research links which aid in the long-term development of the School. Additionally, significant coverage in the national and regional press has been achieved along with television exposure. Benefits arising from this exposure include the attraction of new undergraduate and postgraduate students to the School as they are able to see an exciting project in which students are actively involved in developing a real piece of engineering hardware.

2.5 Benefits to Students

The final year BEng, MEng and MSc students involved in the Formula car project have used increasingly more sophisticated analysis methods as time has progressed. This technology is transferred from year to year, increasing not only the depth of student understanding but the breadth of their experience of advanced analytical techniques. An individual student's enhanced technical exposure is combined with real responsibility for an aspect of the project, a responsibility that can affect the outcome for all the other students. Thus the Formula Car project can be viewed as a small company working to produce a prototype product, the project having all the commercial pressures that would be seen outside of academia. Each student is effectively an employee of that company. Through the course of participating in the Formula Car project, the students develop attributes that are desirable to both industry and the professional institutions. These attributes which are wider than those that a conventional academic project can develop include the following:

- enterprise
- commercial awareness
- business operation
- project management
- team work
- communication
- project data management
- risk analysis
- cost benefit analysis
- benchmarking
- finance

The students learn how to take initiative and gain a feel for what constitutes a performance advantage, not just in terms of outright performance, but also in terms of cost, manufacturability and serviceability. They can see at first hand how a project involving a large number of people should be managed, appreciating the complex interactions between themselves, their more senior colleagues, members of staff (technical, clerical and academic) and outside sponsors and suppliers. An understanding of how to work in a team and how to communicate their ideas to others is gained. The necessity for recording product data in terms of drawings, specifications, suppliers' documentation and experimental and theoretical results is appreciated and a method for doing so is experienced. Through the production of a cost analysis report for the whole car, the use of product data for a specific task is seen and the importance of having a useable system in place is emphasised. Students experience, at first hand, how complex decisions are made and the implications that these decisions can have on the viability of the end result. For example, they often have to decide whether to buy in a non-optimal 'off-the-shelf' component, produce their own part or get an outside manufacturer to make it for them. Each option has different risks both technically in terms, for example, of a part's structural integrity as well as commercially in terms of delivery dates and costs. Students are also exposed to the process of benchmarking, both of individual components and of the overall vehicle performance.

3. MEng AUTOMOTIVE ENGINEERING

The School of Mechanical Engineering at the University of Leeds has a well-established record for excellence in the teaching and learning of core mechanical engineering subjects and advanced topics associated with its principal research areas. Automotive engineering research is a major theme within the School and one for which it has a considerable international reputation in subjects as diverse as computational fluid dynamics, design and manufacture, computer aided engineering, engine combustion, tribology and vehicle dynamics. As a consequence, the School has links with most of the major vehicle, engine, engine component and petrochemical product manufacturers around the world.

Through these industrial links, it has become apparent that the automotive industry has a long term demand for high quality engineering graduates who not only have the prerequisite skills expected of a mechanical engineer but also a more in-depth appreciation and understanding of the specialist needs of the automotive industry. These needs can be considered under three headings:

- application of core mechanical engineering skills to automotive applications, e.g. chassis engineering design

- additional subjects not considered in the mechanical engineering programme, e.g. drivetrain engineering

- methods of management and working methods, e.g. team working.

In response to this need, the School of Mechanical Engineering embarked upon the design of a new undergraduate programme of study in Automotive Engineering. This new programme of study was launched in 1997 with an intake of 11 students, followed by 23 students in 1998. The growth in just 12 months was much greater than had been anticipated.

At the outset, it was decided that the new course would be designed as a four year MEng programme with a three year BEng qualification for weaker students who failed to reach the required standard. The reasons for this were threefold:

- satisfies the need of the automotive industry for high quality graduates

- reflects the imminent changes to accreditation procedures leading to chartered engineer status [2]

- creates valuable time in the curriculum to address transferable skills such as team working and additional academic topics normally only available to MSc students.

In common with all the programmes of study at Leeds, the MEng Automotive Engineering is a modular design based around two semesters. There are a range of compulsory and optional taught modules with a wide range of assessment methods beyond the traditional examination, for example written reports, posters, verbal presentations, design drawings, design and make projects.

Table 1 identifies the key additional features of the MEng Automotive Engineering compared to the MEng Mechanical Engineering programme of study. The first two years are very similar to the Mechanical Engineering programme to facilitate the development of the core skills common to both disciplines. The degree of specialisation, however, increases through the third and fourth years as the emphasis moves to transferable skills and advanced taught modules.

Table 1: Key additional features of MEng Automotive Engineering

Year of Study	Feature
1	• automotive design and manufacture
2	• automotive design and manufacture • automotive engine combustion
3	• vehicle design and analysis • automotive team design • individual automotive project
4	• automotive design team project • automotive industry assignments • individual automotive project • advanced automotive engineering taught modules

Table 2 gives an indication of the range of advanced taught modules available to the students as options in the third and fourth years. To cater specifically for the needs of the motorsport industry, a racecar chassis engineering module has been developed for fourth year students. This applies finite element analysis, computational fluid dynamics and vehicle dynamics analysis to a highly specialised branch of automotive engineering.

Table 2: Indicative optional taught modules for MEng Automotive Engineering

Year of Study	Subjects
3	• vehicle dynamics • engineering computational fluid dynamics • principles of tribology • combustion & thermal energy systems • computer integrated manufacture
4	• racecar chassis engineering • automotive drivetrain engineering • aerodynamics • engine tribology • combustion in engines • competitive product design

Project and assignment work also plays an important role in the final two years of the programme. Third and fourth year individual projects are driven by the needs of industry and often involve direct industrial contribution. For example, recent projects have focused on vehicle crashworthiness, disc brake performance, engine friction modelling and the optimisation of fuel injection systems.

Within the programme there is also the opportunity to study abroad for one year, at an equivalent academic institution in Europe or North America. This is seen as an important option given the global nature of the automotive industry. As part of this year abroad, the students are required to undertake an assignment related to the automotive industry in that country.

No programme of study is ever fixed for any length of time, particularly in the range of optional taught modules offered in the third and fourth years. Future developments of the MEng Automotive Engineering programme at Leeds will include automotive electronics and mechatronics, materials for automotive applications and emissions and environmental issues.

4. MSc(Eng) AUTOMOTIVE ENGINEERING

The role of a traditional MSc in engineering education is currently in a state of revision which is primarily being driven by changes in the EPSRC funding strategy and SARTOR 3 requirements. The nature of the UK industry has also undergone significant change during the last decade: it has moved away from high volume vehicle manufacture to one dominated by automotive component manufacturers and specialist design and consultancy organisations.

The MSc(Eng) Automotive Engineering programme, based in the School of Mechanical Engineering at the University of Leeds, was conceived in the early 1990's. This was part of a School strategy to broaden its extent of MSc provision in an area well resourced through fundamental research and supported by a buoyant UK automotive engineering industry together with an associated demand for high quality graduate engineers. The programme was launched in 1994 with the overall aim of providing graduates with a broad range and depth of knowledge in automotive engineering, and to develop skills in engineering analysis, design, manufacture and management applied to the automotive engineering industry. It is suitable for young graduates who wish to develop the specialist skills relevant to this industry sector or professional engineers already in the industry who wish to use the course as part of a career development programme. It is also suitable as advanced study in preparation for research work in an academic/industrial environment or in a specialist engineering consultancy. The programme is closely related to the School's Formula Student race car effort and was responsible, during the 1995/96 academic session, for setting up the basic theoretical/design framework needed to develop the School's first racecar in the following year. MSc projects, linked to the racecar, continue to run, and these focus on leading the longer term technical development of the car.

The MSc programme is modular, with students taking eight 10 credit taught modules and a 40 credit professional project module. The taught material is drawn from the accumulated knowledge that resides within the School Research Groups and is complemented by core automotive engineering science. The current programme reflects a shift in emphasis towards a high proportion of optional modules with only one taught module together with the Professional Project as compulsory components. This affords students, from an increasingly diverse range of backgrounds, the opportunity to specialise at the outset and so choose a combination of modules that closely match their own perceived future needs or those of their employer.

The racecar chassis engineering module is an excellent example of how the taught course material has developed in recent years. This new module, spawned from the School's involvement with the Formula Student race car, aims to equip students with the overall design principles for racing cars, together with knowledge of chassis structural design, suspension and aerodynamic design. The material is delivered through a blend of conventional lectures, industrial seminars and practical assignments. It is through the assignments that the students are exposed to state of the art finite element analysis, multibody system dynamics and computational fluid dynamics analysis modelling packages, and so gain useful transferable skills.

The growth in IMechE accredited MEng degrees, that will become the basis for chartered engineer status [2], should be seen in a positive light as far as the continuation and development of specialised MSc(Eng) programmes. A correctly structured MSc programme has the potential to provide the matching section needed to upgrade a BEng (Hons) graduate to MEng status. It is worthwhile to note that this upgrading does not have to take place as soon as the BEng (Hons) graduate enters the job market. The MSc also provides a conversion path along which a graduate from a related field of physical science can move to launch a career within the automotive engineering sector of industry. The stand-alone masters degree remains the principal route through which an overseas graduate of engineering can undertake further study at taught postgraduate level.

The Engineering and Physical Sciences Research Council (EPSRC) has traditionally made available a number of Advanced Course Studentships to selected MSc programmes. These covered academic fees as well as maintenance. However, in 1998, the EPSRC announced fundamental changes to its funding model for such courses [3]. In summary, the aforementioned studentships were to be phased out and replaced by Masters Training Packages lasting up to five years. These are to part-fund studentships and support the development costs for specific 'training packages' produced in close collaboration with industrial partners. Thus, the EPSRC is forcing academia to respond to change and recognise that a programme of study has, like any other product, a design life. It is also promoting closer ties with industry by promoting the development of programmes of study that are driven by the needs of industry.

The MSc(Eng) Automotive Engineering programme at Leeds has been a successful undertaking. Modularisation has facilitated the 'generic' course and enabled full advantage to be taken of efficiencies in sharing material with the MEng and, to a lesser extent, the BEng programmes. The development of new modules has allowed individual research groups to document, with high quality notes, a great deal of acquired knowledge that might otherwise have been left shrouded within countless theses. This has led to a valuable resource for the in-house training of new research students to the School. It has also opened up several short course opportunities for a variety of automotive companies. The recruitment of research students into the School and into the wider UK Higher Education sectors has been facilitated by the MSc with approximately 20% of each year's graduates choosing to move into PhD study in the automotive engineering field. The average head count of 12 students per year could be considered low, but class sizes are swelled by in-house research students and fourth year MEng students. Employment prospects continue to be good, with the majority of students accepting offers of employment within the industry well in advance of completing the course.

5. CONCLUSIONS

- The Formula Student race car project has given students exposure to the latest analytical techniques and technological innovations in motor sport engineering. In carrying out a realistic and ambitious engineering project under tight financial and time constraints, the students involved in the race car project have also gained valuable transferable skills which maximise their opportunities in the graduate employment market.

- The new MEng in Automotive Engineering exceeds the expectations of the new SARTOR 3 standard for professional engineering accreditation and, with its strong links to the race car project, offers an exciting alternative to mainstream mechanical engineering which, so far, has proved very attractive to UCAS applicants.

- With its advanced modules and high proportion of related coursework, the taught Masters in Automotive Engineering offers graduate students the opportunity to specialise in this area and the potential to build up the matching sections required to uprate a BEng degree to MEng status. The high quality teaching material developed for this programme can also be made available for MEng programmes, distance learning and short courses for industry.

- These developments have combined to give students following automotive related programmes at Leeds an exciting and rewarding learning experience.

REFERENCES

[1] Williams T D, de Pennington A and Barton D C, 'The frontal impact response of a spaceframe chassis sportscar'. Submitted to IMechE Proceedings Part D, May 1999.

[2] 'The Formation of Mechanical Engineers: The Education Base', IMechE, 1999.

[3] 'Consultation document on EPSRC's support for Masters level training', EPSRC, May 1998.

Postgraduate

C574/019/99

Developing engineers in the automotive industry

P R BULLEN and **P B TAYLOR**
Department of Aerospace, Civil, and Mechanical Engineering, University of Hertfordshire, Hatfield, UK
H MUGHAL
Land Rover Vehicles, Solihull, and Visiting Professor at the University of Hertfordshire, Hatfield, UK

ABSTRACT

The importance of 'business engineers' to the future success of the automotive industry is established. These 'business engineers' require continuing professional development to enable them to take a systems view of the industry and its processes, whilst developing their technical and managerial capabilities. Two programmes of study have been developed to meet these needs. These programmes are developed and run in partnership with the industry and their key features are described together with issues and challenges facing the partnership.

1. HISTORY

The industry has developed over a hundred years from the carriage industry supplying tailor made vehicles for the very rich in the late 19th century, to a global industry using sophisticated technology to produce high quality vehicles for world wide markets.

In the 1910s Henry Ford applied the methodology of mass production to vehicle manufacture and developed the Model T, the first car for the masses. His dream is captured in the following quotation which also illustrates that success was not dependent on mass production alone:

'I will build a motor car for the great multitude. It will be large enough for the family, but small enough for the individual to run and care for. It will be constructed of the best materials, by the best men to be hired, after the simplest designs that modern engineering can devise. But it will be so low in price that no man making a good salary will be unable to own one – and enjoy with his family the blessing of hours of pleasure in God's great open spaces'.

Henry Ford (1907) (1)

In summary Henry Ford's success was due to:
- Technical developments in product design and manufacturing processes
- Organisation of the manufacturing system and utilisation of labour
- Market conditions
- Supply of workers
- Limitations of competitors

The Model T was so successful that at one time it accounted for 60% of all cars in the USA. The mass production system was seen as the only way forward for an industry wishing to produce volume vehicles at low cost but disadvantages began to appear linked with a limited range of products, the quality of the product and the size and structure of the organisation.

Overall improvements in the industry were achieved in the late 1920s by General Motors under the leadership of Alfred Sloan who decentralised the organisation into smaller divisions operating more as separate companies and who also recognised that customers wanted more variety and more sophisticated products. Again problems arose with companies becoming more concerned about making money than about making vehicles that made money. The next major revolution in the industry was brought about by the success of the Japanese and in particular of the Toyota motor company. In the 1950s the production methods of Toyoda and Ohno of Toyota set the foundations of the Japanese motor industry, which from a base of less than 1% of the share in world motor vehicle production in 1955 achieved some 30% by the 1980s, see Figure 1.

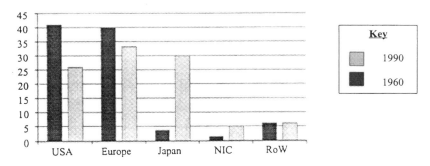

Figure 1 World Car Production- market share %

Circumstances and conditions in Japan were such that modifications to the mass production methods were necessary for Japan to establish a motor industry that could supply the Japanese people. For example, the market was small, requiring a range of products; the workforce was not willing to be treated as a variable cost; and the supply of labour was limited. This led to the adaptation of mass production methods for flexibility and greater worker involvement, which is well documented in the five year M.I.T. international research study, 'The machine that changed the world' (2), where the term 'Lean Production' was introduced. Lean production is presented as a holistic system. It includes design and development, distribution and sales, suppliers and customers, as well as production and is, perhaps, better described by the term the 'Lean Enterprise'. The success of the Japanese industry is well illustrated by reduced lead times to produce a new vehicle and by higher quality vehicles (as measured by the number of defects on the final product per 100 vehicles, traceable to the assembly plant, reported in the first three months of use).

The key points demonstrated by this short history is that successful automotive companies need to match the technology to the human resource and to develop them in harmony with the external factors, with a concentration on adding value to the final product, from a customer perspective.

2. THE INDUSTRY NOW

The automotive industry is now considered to be the world's largest manufacturing industry employing some 15 million people directly and 50 million indirectly with some 53 million new vehicles registered in 1998, world wide. The major companies are located in the major markets of USA, Europe and Japan. The West European car market is the largest in the world with sales reaching over 14 million units in 1998. Figure 2 below shows the pattern of new car registrations in these three major markets and illustrates one of the major problems faced by car companies: the cyclical nature of the market. This is one of the distinctive features of the car industry today. It is a business which combines high technology and fashion, whose final products are heavier and more complex in their process requirements than any other consumer durable. Vehicle manufacturers buy in 60% or more of the value of the final product so that the price and quality of the bought in components have a major influence on competitivity.

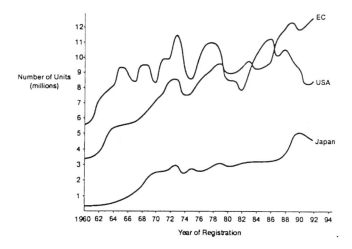

Figure 2 New Car Registrations in the EC, USA and Japan (SMMT data)

The industry needs to be looked at as an integrated whole rather than as a collection of individual companies. Vehicle manufacturers are forming alliances with a smaller number of suppliers, that work closely with them to engineer and manufacture components and modules to the high quality standards required and at acceptable costs. Predictions for the future indicate growth in output but in an atmosphere of intense competition. There is significant over capacity and increasing environmental pressure. Vehicle manufacturers have to produce high quality products at low cost which 'exceed customer expectation' and which meet ever more stringent environmental and safety requirements. The industry is now operating on a global basis requiring a different organisation and outlook in terms of product development and manufacture.

There is much written about the formula for future success in the industry. Very simply the vehicle manufacturer and supply companies are looking to add value to, and reduce the cost of, their products whilst exceeding the expectations of all stakeholders. In order to do this, companies need to employ the appropriate technology and the appropriate organisational structures with a full understanding of the external constraints and most importantly a full understanding of the customer. All of this then needs to be put in the context of increasingly rapid changes of technology, peoples expectations and external factors. John Towers, in a lecture to the Royal Academy of Engineering (3), described the needs of an automotive company: 'You need a design capability of immense technical power. You need an engineering programme which will deliver products to specification and to legislation and a programme which designs cars for ease of manufacture, with all the information coming from a common database to ensure maximum integrity. Then you need a manufacturing capability which is efficient, productive and lean, supported by the intelligent use of manufacturing technology and robotics. Then you need a distribution and dealer network which projects a good image and a quality environment... but most of all you need people, people with a flair to create as well as control the design process. People who understand that engineering is a living dynamic thing... People with a focus on who the customer is... People working as a committed team, using their individual skills and initiative to benefit the process as a whole. The key is continuous improvement. Doing their own job but also finding ways of improving that job.'

The latter part of the quotation places the emphasis firmly on people. It is the education, training and continuing development of these people which will provide the industry with the capability to succeed in the future. This is where Universities have a role to play as a 'supplier' of education and training and as part of the extended enterprise.

3 MEETING THE AUTOMOTIVE INDUSTRY'S NEEDS

3.1 Introduction

The challenge to the University sector is to provide the education and training appropriate to the industry's needs. These needs vary from specialist technical knowledge and skills to more general management skills within the context of the global automotive organisation and the product development process. Automotive engineers therefore need to be able to take a systems view, need to be equipped with the appropriate technology skills and understanding and also to understand the importance of the organisation and human resource. They also need to be able manage and implement change, work in multi-disciplinary and multi-national groups and have the outlook of continuous improvement.

Ian Milburn, Deputy Managing Director of Nissan European Technology Centre, has described the industry's need for 'business engineers' in his keynote address to postgraduate students on the MSc in Automotive Engineering Design, Manufacture and Management (known as IGDS). His 'specification' of 'business engineers' provides a clearer focus. These 'business engineers' are good technical engineers who understand the automotive business. In particular understanding the interrelationship of value to the customer and the cost to produce the product. Value is - customer perceived value. Which will be, for a car, attributes such as performance, quality, delivery, attractiveness and image. Value is determined by engineers from a knowledge of the customers wants. Cost is material plus labour plus overheads. This is determined by time to develop and produce the vehicle, materials and components selected and labour requirements. The key point is that it is the engineer who determines most of the parameters concerning value and cost. It is the 'business' engineers job to creatively 'add value and reduce cost'.

The question now is; how can the University provide professional development programmes to achieve all of this?

3,2 Challenges and key issues

3.2.1 'Customers' and 'products'.
There are many challenges in developing, running and maintaining such programmes of study. The most important driving factors are relevance to industry and appropriate academic standards. The industry will expect delegates (the term delegate is used to differentiate from undergraduate students) to graduate from these programmes with significant 'added value' and significant return 'on investment' especially if they are paying the delegates fees. Similarly the delegates will expect to improve their capabilities within the industry by successfully completing such programmes. Again they will be thinking about the return on their investment of time and money. The University has therefore to offer 'products' which provide this added value and, 'delight the customer', both the delegate and the company.

It is important to relate the structure, within the university, to these 'products' and 'customers'. The industry has moved away from a primary structure based around functions to one based on products with functions playing a supporting role. For example, Ford 2000 has seen a rearrangement of the company into vehicle centres which concentrate on particular ranges of vehicles, for example VC1 is based in Europe and is responsible for the design, development and production of all small and medium size cars world wide. These vehicle centres are supported by specialist functions. A typical university is organised on different lines, normally based on resources and specialisms. Departments are resource bases containing specialist staff and laboratories. Within the department there is even more specialism with individual staff members being specialists within subject areas. A conflict arises as this structure supports the university's strength in a subject, or specialist area, but does not support the customer who is seeking an integrated experience across a range of different specialisms within a particular discipline. This departmental structure and very specialist staff could also provide a poor 'role model' for students seeking a rounded education. This is further magnified when we consider that industry is looking for the 'business engineer', or the engineer described by John Towers with two jobs: their normal job and their job of continuous improvement. One solution is to utilise industrialists alongside academics in the delivery of programmes.

3.2.2 Partnerships with the industry

A key to success is to work closely with industry, as part of the extended enterprise concept, the customer, and partner universities, to ensure the relevance and suitability of the programme and to continually develop and improve it. The philosophy of 'continuous improvement' has just as much relevance to Universities as it does to the automotive industry. Partnership can work at many levels from the strategic to the day-to-day operational level. At the University of Hertfordshire partnership functions at all these levels. The department's Motor Industry Advisory Panel provides strategic direction for all of the department's automotive engineering work. The IGDS (section 4), one of the programmes of study, has a management committee, involving partner companies, universities and delegate representatives, which overseas content, delivery and operation of the programme. Each course within the programme has a module working group, responsible for the detailed syllabus content, its delivery and monitoring and evaluation. The module working group will advise the management committee of the need to update syllabuses and of the outcomes of module evaluation, which always include actions for improvement. This closes the loop and ensures that the quality systems always lead to actions and 'added value' for the delegate. The involvement of senior company managers through delegate mentoring and evaluation, in terms of the increased value of the graduate to the company adds further data to ensure the effectiveness of the programme. Table 1 illustrates the number of automotive companies involved in the IGDS programme and gives a flavour of the range of activities.

Table 1 Industry support for the IGDS programme

	Number of Companies	Number of Individuals
Scheme Management	11	11
Course Development	9	18
Scheme Delivery	35	73
No. of Delegates/Short Course Attendees	27	154
Other Support	8	8
Total	54	

A further dimension to partnership is achieved through associated activities. Companies such as Ford have identified a specific number of universities as 'preferred suppliers' with the intention of improving understanding between partners. Continual evaluation and review ensures the effectiveness of the collaboration. For example, these associated activities at the University of Hertfordshire include research into the applications of computational fluid dynamics to the automotive industry, based partially in the company, and into engineering education and the development of teaching and learning using the new electronics communication technologies.

3.2.3 Flexibility and delivery

Providing continuous professional development (CPD) programmes for industry requires very flexible schemes with equally flexible supporting systems. This involves the structure of the programme, its delivery and issues such as the registration of students. The structure that has been adopted for the programmes described in section 4 is based on short course style modules with large elements of distance learning and 'in-company' projects, so that the delegates are not away from the work place for long periods. The globalisation of the industry, the need for high level education and training by the supply companies and the increasing pressures in the work place requires schemes with even more flexibility. Electronics and communications technologies provide opportunities to meet this demand and offer the potential to enhance the education and learning process and provide supporting systems and information access through an automotive Intranet for example. The experience of conversion of one of the IGDS modules to multimedia, 'electronic learning', format has shown that the development of such systems is very resource intensive in the development stage and that the delegates are somewhat reticent to adapt to the new methods of learning. Lessons can again be learnt from the motor industry. The success of electronic learning will depend on the appropriateness of its use and the training and education of the developer and user. The technological development needs to be matched with the organisation and capability of the human resource.

4. MASTERS PROGRAMMES IN AUTOMOTIVE ENGINEERING

Knowledge of the automotive industry led to the development of a European MSc scheme, in the late 1980s, designed to meet the European automotive industry's needs. This aimed to provide engineers and potential engineering managers with a broad based (systems) overview of the industry and its products together with the developing trends in automotive engineering design, all within a European context. The initial proposals were developed in partnership with the UK and German automotive industries and then included other major European Universities involved in automotive engineering education together with their local automotive industry representatives. The result was an MSc in Automobile Engineering (a European programme integrating technology and management) which has the following key aims and concepts:

- Preparation of engineers for international work
- Study in three countries
- Study in a multinational group
- Study of economic, management and organisational structures and methods in the context of the European automotive industry
- Study of management
- Study to extend technical knowledge in selected areas of vehicle technology
- Study of trends in automotive technology developments

Shortly after the development of this scheme the opportunity arose to bid for funds from the Engineering and Physical Sciences Research Council to establish an IGDS (Integrated Graduate Development Scheme. IGDS is a national framework for a part-time modular Master's level training scheme run as a partnership between business and universities. The overriding objective is to enable students (known as delegates) to perform their jobs better by

continuing their professional development; so the courses are designed not only to impart relevant technical knowledge but also the management skills necessary to make the most of the knowledge. The outcome of the bid was support for the development and start up of an MSc in Automotive Engineering Design, Manufacture and Management (5) offered by a consortium of 5 UK universities, with close links with the automotive industry, which met the following needs of the automotive industry:
- For UK universities to collaborate bringing together their various strengths and specialisation's in a 'unified' way for the provision of graduate automotive engineering education and training

For graduate engineers who:
- understand the industry; its structure and relationships between companies.
- understand the challenges facing the industry.
- understand the processes involved in the successful delivery of competitive products which 'delight the customer'.
- are innovative and creative in the applications of technology.
- are equipped to anticipate and manage change.
- are able to work in multi-disciplinary groups and able to adopt a multi-discipline approach to the development of products

The resulting schemes of study have a similar structure. This is illustrated in Figure 3 for the IGDS.

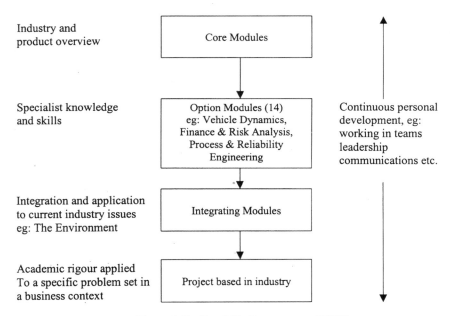

Figure 3 Outline MSc Programme (IGDS)

The novel features include the systems view offered by the core modules, the emphasis on integration of the material and the thread of personal development all coupled to optional modules providing a significant element of specialist knowledge and skills. Table 2 provides an outline content of the core IGDS modules as an illustration of the systems view.

Table 2 IGDS Core module outline content

The International Motor Industry	Automotive Product Development
Social and economic need for land transport	Product life/replacement cycle
The environment of manufacturing	Fundamental and pre-competitive research initiatives
Motor industry infrastructure	Advanced/concept engineering
Corporate strategy	New product programmes
Structure of organisational functions within a company	Manufacturing strategy/simultaneous engineering
Demands of society	Licensing and joint development arrangements
Vehicle life cycle	

These novel features are reinforced by assignments which are linked to the workplace and an examination designed to test the delegates abilities to apply course material in an integrated way. The partnership with industry and other universities is essential to the delivery of the schemes. The first core module, which examines the International Motor Industry is delivered by a 50/50 combination of academics and industrialists and involves some 16 different lecturers together with distance learning material, all co-ordinated by a module leader acting as a 'manager of the learning experience'.

Both MSc schemes are proving very successful. Approximately 25% of the delegates on the European scheme are sponsored by companies such as VW, Mercedes, Karman, and Audi. The delegate mix is multi-national. Delegates on, and graduates from this scheme highlight the benefits of working in international groups and the experience of studying in different countries. They are also appreciative of the input from practising engineers. The scheme has now developed into an MSc in International Automotive Engineering. The IGDS programme only recruits employees from the industry with at least 2 years industrial experience and there are now over 100 delegates registered on the programme, including 70 delegates from the Ford Motor Company. Delegates on this programme are enthusiastic about the involvement of a large number of companies in the programme and the networking that take places. Table 1, above, shows the extent of the partnership with industry for this programme. The programme has been developed to allow maximum flexibility in terms of modes of study and structure of the programme to meet both delegates and sponsoring company needs. A number of new option courses have been developed in response to the advice from the management committee. Flexibility in study is being further increased supported by computer and internet technologies (6).

The graduates from both MSc's believe that they have developed a stronger understanding of the industry and have considerably enhanced their career development prospects.

5. CONCLUSION

- The need for continuing professional development that increases understanding of the industry, its processes and its environment as well as providing increased technical competence and improved managerial skills has been established.
- Two successful programmes of study have been described and evaluated demonstrating how industries educational needs are being met.
- The need for flexibility in delivery and for close partnership with the industry has been highlighted

REFERENCES

1. Lacey, R. (1986) Ford The Men and The Machine. New York:Random House Inc.

2. Womack, J.P., Jones, D.T. and Roos, D. (1990) The Machine that Changed the World. London and New York: Rawson Associates/ Macmillan.

3. Towers, J., 'The emergence of Rover', Engineering Manufacturing Forum Lecture 1994, The Royal Academy of Engineering, London.

4. Bullen, P.R., Jackson, A. and Watson, D.G., 'European collaboration in the education of undergraduate and postgraduate vehicle engineers', Proceedings of I.Mech.E. Autotech Conference 1993.

5. Bullen, P.R., Jackson, A. and Taylor, P., 'University- Industry Parterships providing postgraduate schemes for the automotive industry', ASME Mechanical Engineering Depts Heads Education Conference, March 1997, San Diego, 'Mechanical Engineering for Global Practice'.

6. Bullen, P.R., Chatterton, P. and Taylor, P., 'The Virtual Automotive Learning Environment', IMechE, Education in Engineering at Autotech '99, November 1999.

C574/024/99

The role of research in learning and personal development for engineering excellence in the automotive industry

A J DAY and **R S F HARDING**
Department of Mechanical and Medical Engineering, University of Bradford, UK
K W MORTIMER
Ford of Europe, Marlborough, UK

ABSTRACT

Research is still often viewed as the pursuit of abstruse scientific knowledge, associated with theoretical advance and distanced from practical application. In this paper the role of postgraduate research in learning and personal / professional development is discussed. Two case studies are presented from a Research Masters programme at the University of Bradford which has enabled nearly 80 Professional Engineers to extend their education, continue their professional development, and identify significant savings to their employers. Conclusions are drawn which indicate that Research Masters programmes are flexible, timely and effective in delivering postgraduate qualified engineers for the modern Automotive industry.

1. INTRODUCTION

Industry needs research to establish and maintain its competitive edge, not only by providing new products, but also by solving problems and identifying potential improvements in both products and processes. Despite the business benefits which have accrued from the application of research in engineering design and manufacture, University academic research is still too often viewed as the pursuit of abstruse scientific knowledge associated with theoretical advance and not immediately associated with application. Industrial research (as the name suggests) is often used as a term to describe research which is carried out in industry, as opposed to academic research which is carried out in Universities: the former is "applied" whereas the latter is "fundamental". Although learning is recognized as an essential outcome of any research, it is regarded as specialized and narrow. The holders of "research" degrees, though undoubtedly highly intelligent and educated, are "experts" only in a limited field. Any use of their knowledge or learning outside that particular field has been seen as minimal.

The role of teaching in learning has, in recent years, been under review. Experience, learning through discovery, and "open" learning are a few "new" approaches to learning which are essentially research-based learning albeit at an elementary level. The teacher, instead of adopting a didactic approach, assumes the role of tutor, to advise the student when required. The advantage claimed for such methods include better learning motivation and self-pacing. This paper sets out to show how research ("careful search or enquiry", "endeavour to discover new or collate old facts, etc.", "course of critical investigation" - Oxford English Dictionary) - specifically in the Engineering discipline - offers an increasingly important educational vehicle.

At the academic level of an undergraduate degree, research has become accepted as having something to offer in the way of learning, largely in the form of project work but often still subservient to the taught element with its associated formal examinations. At Masters level, courses are traditionally taught programmes (1) with a project element based on, and following, the taught element. At the higher academic level of the PhD. degree, research is traditionally the driver, but recently the role of a taught component in the PhD. degree has been highlighted in, e.g., the Engineering Doctorate scheme (2). Such a scheme represents a significant advance in the understanding and appreciation of high level learning and its application by the engineering industry, and reflects some of the standard academic practice in North America.

The time required to complete the degree programme is important to industry. Three years is considered by the majority of industry to be simply too long to wait for either the research or the high level learning to be completed. Industry works on shorter timescales and is looking for a shorter-term alternative. For this reason considerable interest has developed in the Research Masters (MRes) degree format; essentially a research - lead Masters degree course with an associated generic taught element representing the "core competencies" required of a postgraduate level qualified engineer. Another advantage of the Research Masters approach is that it is inherently flexible: learning can be tailored to meet the need.

2. THE BRADFORD RESEARCH MASTERS PROGRAMME

The University of Bradford has been operating a Research Masters degree programme since 1994, as part of the Masters portfolio in the School of Engineering. It was founded upon the operating department's excellence in engineering research, its understanding of the generic skills required for postgraduate qualified engineering staff, and its experience with post-experience external students. Quality Improvement was the identified driver for higher levels of education in areas of the Automotive Industry, and the programme was initiated as a partnership with the Ford Motor Company and its Supply Chain. Since starting in 1994, the Research Masters programme in Engineering Quality Improvement at the University of Bradford has successfully enabled 39 Professional Engineers to extend their education, continue their professional development, and identify massive savings to their employers through research based on the principles of Engineering Quality Improvement. Another 40 Engineers are still working on their MSc. degrees, and each year approximately 12 Engineers enter the programme from the automotive industry.

The Masters programme is a project-based structured research degree programme, which integrates the benefits of academic rigour with the practical working initiatives of the sponsoring company. It provides the opportunity for innovation and personal development aligned to Quality systems, and emphasizes the culture of the learning organisation. It has proved to be a programme which addresses the needs of modern industry, integrating work-based learning with research project applications, and providing flexible, tailored, learning opportunities using key methodologies. It brings together Industry and Academic expertise, and emphasises the Customer - Supplier relationship. Its potential has been recognized elsewhere in the world in the form of a collaboration with Michigan Technological University in North America, while other Universities in Australia, South America and Europe are developing similar research-based structured Masters programmes.

The Masters programme aims to develop research, technical and motivational skills, to complement and consolidate the understanding and application of methodologies for continuous improvement, to generate familiarity and understanding of Robust Engineering Design techniques, and to bring academic standards of research into work-based learning. The learning outcomes include:

- Competence and confidence in the understanding and application of the disciplines of Quality Engineering, including Failure Mode and Effects analysis (FMEA);
- Transferable skills in Experimentation, Process Management, structured problem solving, Quality Function Deployment, all with the associated people skills;
- Transferable skills in modern methods and procedures for data collection, modelling and analysis;
- Core skills in using and understanding techniques and methods for customer and other data collection through surveys and questionnaires;
- Core skills in researching and evaluating existing work both within the Company and in the public domain;
- An academic thesis which satisfies the rigorous academic requirements of the examiners and the expectations of the Company.

The programme framework is based on an individually tailored programme of learning modules to prepare for and support the research:

- Introduction to research methods and study skills, continuing research skills development schools;
- Quality Engineering modules including Total Quality Management, Statistical and Probability methods, Data processing and analysis, Statistical Design of Experiments, Structured problem solving, Statistical Process Control, Quality Function Deployment, Reliability, FMEA, Robust Design.
- Engineering specialisms for the chosen research project, drawn from the University's MSc. programmes, e.g.: Design for Manufacture, Engineering Materials, Finite Element Analysis, Structural Dynamics, Numerical and Computational methods.

The student must write up and complete a Master's Thesis which must be submitted and pass a formal viva voce examination by the External and Internal Examiners, and a formal presentation at a company-based conference. In the former respect each successful candidate gains a Research Masters degree - a "mini- PhD".

CASE STUDY 1: "IMPROVING VEHICLE SAFETY BY LISTENING TO THE VOICE OF THE CUSTOMER"

This research project (3) was completed by a degree qualified Mechanical Engineer with some 10 years experience in a specialist role in the automotive industry. He then moved into the connected but relatively new area for him of vehicle safety, and one of his highest priorities was to use the literature study part of the research to build up his subject knowledge.

A typical dictionary definition of safety is "freedom from danger"; a flawed definition because almost every human activity involves some element of risk. The British Standards Institute (BSI) definition is more realistic: "freedom from unacceptable risks of personal harm", which considers safety as a "subjective" quality which depends upon what a particular individual, organization or society regards as "acceptable". This Masters research project had the personal objective for the student to understand the factors which influence the concepts of safety as a subjective quality, while for the sponsoring company the objective was to understand how technical advances in the science of motor vehicle safety could be best applied to the vehicles it designs, manufactures and sells for use on public roads. The subject bridged several disciplines and there was no taught course at Masters (or indeed any level) which addressed the needs. The sponsoring company viewed the outcomes as being urgently required for strategic reasons, relating to the introduction of the European New Car Assessment Programme (NCAP).

The principles of Total Quality Management (TQM) include the definition of the term "quality" in terms of the consumer's needs and expectations. The term "voice of the customer" is used to describe the disciplined approach to obtaining, understanding and prioritizing customer requirements, and this approach was adopted as the methodology for the research. However, a major obstacle in identifying voices of the customer in vehicle safety is that most consumers, viz. vehicle occupants, are not involved in accidents which result in injury. As an example, of the 25 million vehicles registered in Great Britain in 1995, only 230000 injury accidents (involving 415000 vehicles and 311000 casualties (4) were recorded, a ratio of 1 in 100 vehicles approximately. In the United States a crash occurs once in every 335000 driver kilometres or every 16 driver years of driving. Fatal accidents occur very infrequently; once every 93 million driver kilometres or once every 4300 driver years of driving.

As a result, most consumers of the products of the motor industry do not have first hand experience of how their vehicle would perform or how well protected they would be in a crash, and therefore cannot evaluate their level of satisfaction. An initial conclusion of the research was therefore that the voice of the customer in vehicle safety is silent. This is very different from other features of the vehicle such as fuel consumption or acceleration where every consumer has personal experience on which they can comment and determine their level of satisfaction.

Many surveys have indicated the high importance that customers attach to safety, e.g. in a list of the improvements most wanted by customers where, worldwide, safety in accidents came second to fuel economy (5). However, such surveys are not sufficient: they present a snapshot in time, they are concerned with features and not performance levels (i.e. having an airbag is a desirable feature in itself, but the reduced risk of head or chest injuries is not quantified), and they are concerned with available technology only.

Direct consumer feedback can come from satisfied customers who write to the vehicle manufacturers after having been involved in an accident, which theoretically represents the voice of the customer. However, it cannot be assumed that such customers are a representative sample of the accident population as a whole. Dissatisfied customers also write to complain about their vehicle's safety performance after a crash, but these are dealt with through the formal channels of Product Liability. Such feedback does represent the voice of the customer, but the seriousness of such complaints and the formal procedures by which they must be treated takes precedence.

Actions associated with such complaints aim to bring a particular product up to expected standards rather than to set targets for improved safety standards.

Indirect customer feedback can come from organisations who represent customers in the areas of vehicle safety, and from experts who have studied real world accidents and analysed the data, reported on it indirectly, and drawn systematic conclusions.

Based on this survey of the possible sources and types of consumer feedback the concept of the "surrogate" voice of the customer was developed. This included expert feedback from real-world accidents, laboratory crash tests, the use of instrumented crash dummies, safety rating scales, the abbreviated injury scale (AIS), the head injury criterion (HIC), and the New Car Assessment Programme (NCAP). The scientific basis of vehicle safety rating scales is empirically founded, relating, for example, the risk of serious chest injury to measured dummy chest deceleration values through the equation:

$$P_{chest} = (1 + \exp(5.55 - 0.0693 \times \text{chest deceleration}))^{-1}$$

From such empirical bases, the risk of life-threatening injuries can be evaluated: that used in the US NCAP is shown in Figure 1. This indicates the combined probability of a "severe" threat to life from injuries received, perhaps equivalent to serious, permanent injury and debilitation, and effectively represents what the effect on the customer would be in the event of a crash - a "surrogate" voice of the customer. This knowledge can feed back into the design process so that predicted or measured chest and head decelerations can be combined to provide the required level of protection to the consumer in the event of an accident.

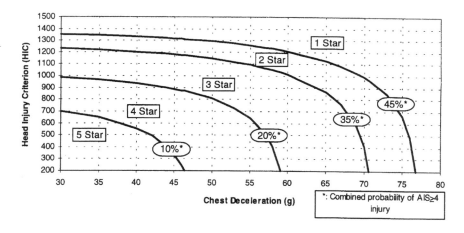

Figure 1   Injury risk evaluation used in US-NCAP

The outcomes of the research were that two "surrogate" voices of the customer were potentially useful in improving motor vehicle safety by design. These were the various predictive

safety rating systems operated by bodies representing consumers and the data coming from in-depth investigations of real world accidents carried out by expert investigators. The availability of such surrogate voices of the customer made up for the limitations of the available direct consumer feedback in the field of vehicle safety design. A structured approach to using these data was proposed:
- Development of new safety strategies,
- Identification of safety defects,
- Identification of vehicle design details which could be improved by either minor or major engineering changes.

A model for incorporating real world safety experience into vehicle design was developed as shown in Figure 2:

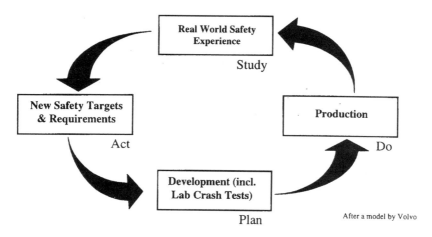

Figure 2   model for incorporating real world safety experience into vehicle design

This model, which reflects the Shewhart cycle of Plan, Do, Study, Act (PDSA), enabled the sponsoring company to formalize its use of the two "surrogate" voices of the customer (predictive safety rating systems and data from real world accidents) in the continuous improvement of its products. As real world experience with that vehicle is gained (which can only be gained over time), and new safety target and legislation evolves, the design cycle is appropriately informed.

Potential areas for improved safety design were also identified, including:
- Areas where improvements have been made but the effects cannot yet be evaluated for the real world, e.g. structures for side and offset frontal impact.

- Areas where injuries occur but the mechanisms are not fully understood, e.g. injuries sustained by non-struck-side occupants. This was a new finding of this research, not previously recognized.
- Areas where the injury mechanisms are understood but no suitable solution has yet been developed, e.g. leg injuries in side impacts.
- Areas where injury mechanisms are partially understood, e.g. leg injuries in frontal impact,
- Areas where safety features are not proving effective because of occupant behaviour, e.g. refusal to wear seat belts.

The benefits to the company sponsor from this research came in terms of establishing safety performance specifications for new vehicle design. Some Automotive Manufacturers market their cars on the basis of exceptional safety, while other makes of cars perform well in tests such as the Euro-NCAP programme, but are often not perceived to be "as safe". The research outcomes included developing methods for translating the output from safety studies into engineering targets, thereby allowing the use of such data for identifying engineering changes (both major and minor) which will improve real world safety performance. Of particular importance is to carry out further analysis into the injuries sustained by non-stuck side occupants in side impact, and into leg injuries in side impact. The cost benefit to the Company lies in greater safety and improved sales.

CASE STUDY 2: "EVALUATION OF COATINGS FOR USE ON GLASS RUN SEALING SYSTEMS"

This research project (6) was completed by a diploma qualified Polymer Technologist with some 8 years experience in a laboratory role in the automotive supply industry. As part of his planned career development he moved into the job of Project Engineer, with responsibility for new vehicle glass run sealing systems. His objective was to use his considerable technical knowledge formally in designing and developing new systems and products.

Manufacturers' and customers' expectations of the performance of the sealing systems around the closures (doors, windows, and boot/hatchback) of motor cars have increased for reasons of styling and appearance, minimising wind noise, and durability. The motor car's aesthetic design often impinges on the functional performance of seals because of shape complexity, materials, processes of assembly and cosmetic appearance. The student aimed to understand the scientific, technological and customer factors which influence the vehicle closure sealing system design, from materials through processing to in-service performance and durability with particular reference to the replacement of traditional rubber materials with new polymeric materials. In particular the use of polyurethane coatings in window glass run seals was to be investigated for reasons of improved appearance, including colour co-ordination, reduced cost, and improved durability. For the sponsoring company the objective was to prove out technical advances prior to bringing a new product to market.

Ethylene Propylene Diene Monomer (EPDM) has been used in automotive window glass run systems since the 1960's. In order to reduce the window winding force, reduce the occurrence of glass marking and operational noise and judder, and to improve the durability of the glass run, a "flock" coating has conventionally been applied to the base EPDM seal (flock comprises short fibres of nylon or polyester bonded on end to the base seal). Apart from the cost of the EPDM / flock system, its tendency to hold moisture can cause damage in freezing weather conditions.

Cosmetic polyurethane coatings are extensively used on body components such as bumpers and spoilers. They have the potential of reducing the glass / seal friction, improving durability, avoiding freezing damage, and reducing wind noise. Reduction of the glass / seal friction offers a potential cost saving because less power is required to open or close the windows, and therefore a smaller electric motor can be specified.

The physics of polymer and elastomer friction and wear were studied to provide an understanding of the seal behaviour when sliding against glass. A detailed study of the chemistry and processing of polymers and elastomers, including polyurethane chemistry, provided the scientific base on which to develop a research methodology in the required timescale. It was established that the properties of polyurethane can be tailored to suit any application. In particular, for glass runs, additives such as silicone oil, silica, PTFE and Molybdenum Disulphide can substantially improve abrasion resistance and reduce friction levels.

As a result of these initial studies, a polyurethane coating formulation was identified as having potential for glass run seals, and a base of EPDM was selected because of its established properties of flexibility and durability. The research then moved on to methods of proving out the optimisation of the glass run seal with these components in actual operation.

The sponsoring company was a major supplier of glass run systems to the automotive industry, and study of the associated failure modes confirmed the need to take into account the end user's perception of failure. These included smearing of the glass by the rubber, and uneven operation associated with judder and noise during window operation. Conventional development and durability tests were found to be based upon "Engineering Specification" tests from the automotive manufacturer being supplied which simply required that the system performed within some specification to a minimum number of cycles. Such tests were not representative of end-user perception of failure.

A "Quality Engineering" methodology (7) was developed, starting with the definition of the "ideal function" associated with the glass run seal. This was defined as "to seal the glass to the door of the vehicle and allow smooth travel when raised or lowered". Error states included water leakage, draughts, wind noise, judder and marking of the glass. A statistically designed experiment was then prepared to investigate the functional performance of the glass run seal over time - the durability. This required the identification of a "quality characteristic" which was measurable, consistent, and representative of the functional performance.

The chosen characteristic was the power required for the motor to raise the window glass. The experimental design matrix is shown in figure 3:

	CONTROL FACTORS						NOISE FACTORS
			Large	Small	Large	Small	WINDOW CHANNEL WIDTH
COATING THICKNESS (microns)	PRIMER/ NO PRIMER	EXTRUSION BASE THICKNESS (mm)	Large	Large	Small	Small	WINDOW END FLOAT
15	No	1.5					
15	Yes	2.5					
25	No	2.5					
25	Yes	1.5					

Figure 3   **Experimental design array**

The control factors were:
- Coating thickness,
- Extrusion base thickness,
- Primer / no primer.

The noise factors (uncontrollable variation) were:
- Window channel width,
- Window end float.

A test rig was designed and built, comprising a passenger car door with adjustable glass run seals and a supply of electric winder motors to counteract motor deterioration (each run in the experiment lasted 30000 cycles of window winding and the motor was replaced prior to each run). Production variation was minimized by close attention to processes; all the EPDM base extrudate was manufactured in the same batch, and preparation for the coating was meticulously controlled.

The results were logged for each cycle, and the average power was calculated at 5000 cycle intervals up to 30000 cycles. It was clear that for some of the formulations the average power required to wind the window increased with time, proving deterioration over time, as shown in figure 4. Eventual failure was associated with a rapid increase in average power, as indicated in (Acton).

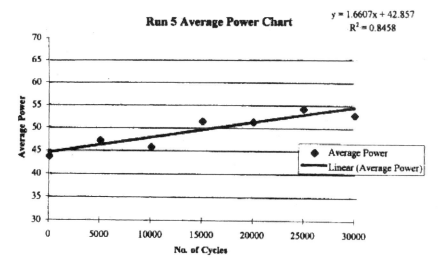

Figure 4   Average power vs. time for test run 5

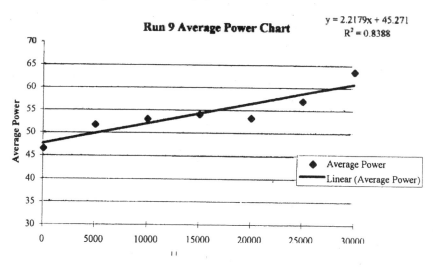

Figure 5   Average power vs. time for test run 9

Analysis of the results indicated that the factors which had most influence on the average power were the thickness of the extrusion base, and an interaction between the thickness of the extrusion base and the channel width. A thicker extrusion base was found to be better; it was surmised that this affected the bulk deformation behaviour of the material supporting the coating, and the extra thickness would contribute to the ability of the glass run to be resistant to abrasion and wear. The interaction between the thickness of the extrusion base and the channel width was such that a greater channel width and a thicker extrusion base increased the seal durability; this could be attributed to the beneficial effect of the thicker extrusion base and the reduced normal force at the seal/glass interface generated by a seal held in a thicker channel.

The outcomes of this research enabled the sponsoring company to commence production of a polyurethane coated glass run sealing system with an optimal design (in terms of extrusion base thickness) for durability. The window channel width was recommended to be as wide as possible without being unacceptable in terms of other performance criteria. This generated a design which was durable and "Robust", to variations in window channel width and end float, both representing uncontrollable variation arising from the production tolerances of the pressed steel car door and window frame. A surprising outcome was that neither the coating thickness nor the presence of primer had a significant effect on the system durability, because clearly as the coating thickness tended to zero the resistance to wear would be reduced. This provoked much critical discussion, and it was concluded that the system was insensitive to those factors around the levels that they were set at, so that a "Robust Design" had been arrived at through specifying the "technically best" polyurethane coating/base material combination, and the thickness. The company's technical knowledge and understanding of polyurethane coatings on automotive glass run seals was considerably enhanced, enabling them to bring the new product to market. This lower cost option to flock coating could thus be used more extensively and improve competitiveness by reducing costs.

DISCUSSION AND CONCLUSIONS

The two case studies shown above have been chosen from 39 research projects completed under the Masters research programme, to represent two very different types of research, learning and specific outcomes. In both cases the overall outcomes satisfied all the personal and professional objectives set, generated knowledge and expertise in the individual through project driven learning, and delivered product and process improvements to the sponsoring company.

The formal module component of the Masters programme was designed to be "generic", i.e. the learning gained related directly to each individual research project in terms of the tools, techniques and methodologies applied. The Students not only learnt about such fundamentally important methodologies such as statistically designed experimentation and robust design, but also applied it for real. This gives them the confidence to use such formal techniques in future projects so that "right-first-time" becomes a practical reality. The overall programme deliberately has a Quality Engineering theme since this is fundamental to the "customer-driven" philosophy which is fundamentally important to the modern Automotive Industry.

This approach to learning has been found to be appropriate for many professional Engineers in the Design, Manufacturing and Research areas of the sponsoring companies. Maturity has been found to be an essential entry qualification, and there must be a foundation of formal education equivalent to a first degree. The students must be able to balance the conflicting demands of work, domestic and project pressures, and must be able to take a broad view so that the research is genuinely conducted to build on, and advance from, existing knowledge.

The projects were both closely job-related, and the Students learnt extensively about the science associated with their job function. In each case the academic supervisors had some specific knowledge in closely allied subject areas which enabled them to guide the research correctly. The Students learnt how to apply proven scientific research and analysis methods, and by so doing, developed a deep understanding of the theoretical basis of such methods. The rigour of following a research programme, keeping detailed records of the study, and writing it up formally as a research thesis, has developed critical faculties which have been found by the students and the sponsoring companies to equip them well for senior roles in the industry. The flexible and focussed approach to learning has satisfied the sponsoring companies in delivering expertise in strategically important areas of business.

REFERENCES

1. "A knowledge-based system for postgraduate engineering courses", A.J. Day, A.K. Suri, Journal of Computer Assisted Learning, Blackwell Science ltd., 1999.
2. "An introduction to the EngD", Engineering and Physical Sciences Research Council booklet, Swindon, U.K., July 1999.
3. Improving vehicle safety by listening to the voice of the customer", P.A. Fay, MSc. thesis, University of Bradford, U.K. 1997.
4. "Road Accidents Great Britain 1995 – the Casualty Report", Department of Transport, HMSO, London 1996.
5. "Global Image Study", Market Opinion Research, Farmington Hills, USA, 1995
6. "Evaluation of coatings for use on glass run sealing channels", R.M. Acton, MSc. thesis, University of Bradford, U.K. 1998.
7. "Quality Engineering using Robust Design", M.S. Phadke, Prentice Hall, 1989.

C574/037/99

Development of a Master of Science programme in automotive systems engineering

S J WALSH, A MALALASEKERA, and **T J GORDON**
Department of Aeronautical and Automotive Engineering, Loughborough University, UK

SYNOPSIS

Starting in 1986, in partnership with the Ford Motor Company, Loughborough University has been working closely with the automotive industry in designing, developing and delivering a part-time MSc programme in Advanced Automotive Engineering. Recently the entire programme has been reorganised to incorporate a theme of Systems Engineering. The new course retains a strong automotive engineering content but with an added element of 'top down' design. This paper explains the motivation behind the new course and includes a discussion of recent experience using a newly developed Computer Based Study Support System.

1 INTRODUCTION

The Loughborough MSc programme in Automotive Systems Engineering is aimed at engineers working in the automotive industry, it is not designed to train engineers for their daily job in industry. Rather, it seeks to widen horizons and build confidence, providing a route for MSc graduates to acquire:
- knowledge and technical expertise in a wide range of automotive disciplines
- a systems viewpoint for automotive design and manufacture, with specific skills in formulating automotive engineering systems in terms of their function and performance
- relevant and in-depth knowledge in chosen areas, through elective modules
- the ability to transfer new skills and knowledge to the workplace, via the industry-based MSc project
- a confident and open-minded attitude to exploring new areas of knowledge in the future

The programme is aimed primarily at product development engineers, however, through the correct choice of elective modules the programme offers considerable depth in manufacturing engineering.

In response to industry's need for more work place based study a distance learning element has been introduced into the course. In parallel a computer based study support system has been developed and implemented to assist the student in studying away from the university. This paper includes a description of the support system.

2   AUTOMOTIVE SYSTEMS ENGINEERING

Systems Engineering can mean many different things to many different people. In this MSc programme it is considered to be a pragmatic way of dealing with the complex products, processes and constraints that fill the automotive world. Gone are the days when engineers might seek to build 'good motor vehicles' by the simple strategy of specifying 'good quality components'. Designing and building with confidence involves quantifying the function and performance of systems and sub-systems. 'Good' engineering practice is still needed, but applied in a way that ultimately links the bottom level component design to the top-level objectives such as customer satisfaction and cost effectiveness. No engineer working in the modern automotive industry can afford to ignore this functional approach.

To reflect this, the programme:
- adopts a top-down approach to the delivery of the vehicle engineering topics
- incorporates a 'systems thinking' framework, referring to product lifecycle, target setting, requirements capture and cascade, plus elements of business-related drivers for engineering practice
- includes a very significant level of core technical engineering content
- emphasises the duality of approach to engineering: components and assemblies *versus* functional systems, physical *versus* functional attributes and boundaries, etc., and starts to develop these themes at the vehicle level
- provides clear links between design and manufacture, for example presenting examples where manufacturing capabilities have a large impact on design and system robustness

3   PROGRAMME FORMAT

The MSc comprises 180 modular credits, made up from eight taught modules valued at 15 credits each, plus a Masters Project valued at 60 credits. The programme is designed for part-time study by graduate engineers working in the automotive industry. Of the eight taught modules, four are designated as being 'core' and four more 'electives' are chosen from a list of available options. The core modules are normally studied during the first academic year of the programme, and the electives during the second year. The project is initiated towards the end of the second year, with completion in time for graduation in the summer of the third year. However, in recognition of the sometimes unpredictable demands of the students' work and other commitments, there is considerable flexibility in the time available to complete the programme, and it is usually possible to defer taking any particular module. Table 1 summarises the modular structure of the programme.

Each 15 credit taught module is designed to occupy approximately 10 weeks of part-time study. Central to each module is an intensive week of residential study at Loughborough,. Outside of this, students are expected to work on prepared course materials and assignments, and to do this effectively, they are supported by an on-line delivery system. An interactive 'discussion database' is used to provide additional information and materials, and also to facilitate tutorial-type discussion. This computer based system is password-protected, and accessible by students using a standard web browser such as Netscape.

3.1 The core modules

The Core Modules provide a broad foundation for understanding the wider aspects of automotive engineering. The first two core modules work from a vehicle engineering perspective, developing the relevant engineering fundamentals alongside a top-down review and analysis of the major vehicle systems. This perspective is then widened to deal with more general Systems Engineering concepts and methods. Finally the link to manufacturing and simultaneous engineering is explored. A list of the core modules is given below:

- Engineering Framework
- Vehicle Systems
- Systems Engineering
- Manufacturing Systems

'Engineering Framework' and 'Vehicle Systems' are pitched at the level of vehicle systems and attributes, including elements of the 'top down' focus, particularly in the area of customer and legislative requirements, and how these apply to the major vehicle functions such as straight-line performance, fuel economy or vehicle dynamics. There is also a very strong emphasis on developing the accompanying engineering tools and concepts. The focus is widened in the next two modules 'Systems Engineering' and 'Manufacturing Systems'. These modules address key areas that interface with vehicle design, such as:

- **vehicle life and lifecycle** - where an extended time-line for the vehicle is taken. Thus, issues such as product usability (ergonomics), reliability, recycling, maintainability are dealt with. Also considered is the interaction of vehicle systems with other systems. For example, telematics and highway information systems. An extended list of requirements is thereby generated.

- **vehicle design as a process** - here the methods and procedures relevant to a systems-based vehicle design process are considered. This includes relevant systems engineering tools and methodologies, such as requirements capture and cascade. An overview of CAE testing and sign-off practices also comes under this heading. The link to manufacturing is also explored via issues of simultaneous engineering practice and the need for design for manufacture and design for assembly.

- **vehicle engineering as a business** - the links to commercial and economic factors are briefly explored. This includes cost implications and planning for the design, manufacture, service and disposal phases of the vehicle lifecycle, as well as associated issues of vehicle programme timing. This is an area of common concern for the design and manufacturing areas of the automotive industry.

- **manufacturing systems** - here manufacturing is presented in a way that mirrors the vehicle engineering content of earlier core modules. The emphasis is on manufacturing processes and their organisation, cost, effectiveness, capabilities and limitations.

3.2 The elective modules

In contrast, the remainder of the MSc programme provides an opportunity to add considerable depth, first through the elective modules, and secondly through the MSc project. The electives cover areas such as powertrain design, vehicle dynamics, manufacturing and materials. Though the Systems Engineering aspects are less explicit here, it will always be a priority to highlight the relevance and significance of the technical material covered. A list of the elective modules is given below:

- Powertrain Engineering
- Vehicle Platform Engineering
- Engine Performance and Design
- Vehicle Dynamics
- Vehicle NVH
- Manufacturing and Materials Processing
- Automotive Control
- Design Integrated Manufacture
- Advanced Automotive Materials

3.3 The masters project

Not surprisingly, the project is also expected to provide this same blend of systems engineering framework plus detailed technical engineering content. Typically this is carried out at a company location, under the supervision of a nominated University supervisor, and with the co-operation of a company-based manager or supervisor. The project topic is agreed at an early stage between the student and the academic and industry supervisors, and regular discussions take place. A good project combines the academic rigours of the university with the technical and commercial requirements of the company. Project dissertations can have access restrictions where commercial confidentiality is an issue.

4 MODULE DELIVERY

As mentioned previously, the MSc in Automotive Systems Engineering is aimed primarily at engineers working in industry. In the first year there are four core modules to be studied and in the second year four elective modules. Each of these taught modules occupies a ten week study period centred on a one intensive week of residential study at Loughborough University. The study period begins three weeks prior to the residential week and continues for six weeks after the end of the study at Loughborough. The structure of this study period is shown diagrammatically in Figure 1.

The pre- and post residential study period is supported by the operation of a Computer Based Study Support System (CBSSS). Approximately 6 weeks before the residential week the student receives a mail with brief instructions referring them to the CBSSS. The CBSSS provides a module guide, module overview, key dates, lecture topics, staff contacts, assessment details together with details of the pre-residential study. The nature of the pre-residential study will vary from module to module, but is typically equivalent to nine hours of classroom contact - perhaps 6 lectures and 3 tutorials. It may consist of background reading with structured exercises, computer-based assignments (for example, computer modelling) or whatever the lecturer feels is most appropriate. Required pre-residential study materials are made available at the same time either in electronic form on the CBSSS, or as hard copy printed materials via the normal post.

During the residential week itself there will be a mixture of lectures, tutorials and laboratory sessions. Though the week is quite intense the activities are structured so as to make the time both interesting and challenging. After the residential week there is typically the equivalent of 6 hours of classroom material to be studied as follow-up assignments. There will also be assessed courseworks to complete.

The actual study hours are very hard to quantify, since different people work at different rates and with different levels of commitment, so the numbers given in Figure 1 are intended as an indication of our expectations based on current experience. This might typically be ten hours of personal study per week for the pre-residential study, with 10 hours per week over the following seven weeks being devoted to the post residential study and the coursework assignments. Obviously the workload might not be spread so evenly in practice - some students tend to concentrate their efforts during a shorter period close to the hand-in deadline!

4.1 The computer based study support system

The Computer Based Study Support System (CBSSS) was introduced to assist with the reduction of the residential period at Loughborough from two weeks to one week for each MSc taught module. The main objectives of the CBSSS are:
- to deliver study support material efficiently at a distance
- to provide an effective communication system between the teaching staff at Loughborough University and the students studying at the work place

Several issues were considered in designing and developing the CBSSS. For the University, the system should:
- be easy to develop and to modify
- be cheap to run and maintain
- provide access to students from the UK as well as abroad since the prospective students are local, European and International
- not involve a high level of IT expertise in developing and maintaining

For the users, the system should be:
- efficient and user friendly
- suitable to run on computers with low technical specifications
- easily accessible
- able to run on most available hardware platforms
- able to run with freely available software

It was decided to use the World Wide Web (WWW) as the delivery platform since it is easily available. Lotus Notes R4.5 was chosen as the WWW development software due to its powerful in-built features. Adobe Acrobat Portable Document Format (PDF) was found to be a suitable form to create electronic documents due to its cross platform nature and also due to the compact files generated. PDF documents also maintain the look and the layout of the original documents. In order to read the PDF documents, users need Adobe Acrobat Reader software installed in their machines. Adobe Acrobat Reader software is currently available to download from the Adobe Acrobat Web site free of charge.

4.2 The use of the CBSSS

The CBSSS is a password protected WWW site and it provides information and necessary study support for students studying at the work place. A screen dump of the homepage is shown in Figure 2. The list of available areas in the CBSSS can be seen in Figure 2. Documents under some of these areas are sub-categorised according to appropriate 'Team' names or module names. Each MSc student intake is identified using the 'Team' name. The C Team is this year's intake and the B Team refers to last year's intake. The CBSSS is developed in a way that the new and modified documents can be identified easily by the appearance of the 'New' or 'modified' icons in front of the document title in the content pages. These icons will appear automatically if the document has been created or modified within the last four days.

The first item in the available list, the 'What's New/Modified' hyperlink, shows all the new and modified documents in the CBSSS except in 'Questions & Answers' and 'Chat for Students' areas. The reason that these two areas are excluded is because in these areas students and staff create their own documents. Currently, the time limit for the automatic appearance of the new and modified documents in the 'What's New/Modified' area has been set to two days. This time limit might need to increase after considering the feed back from the staff and students.

The 'Noticeboard' is the area for general information and it contains notices with technical and non-technical matters. 'Library Services' is a hyperlink to the 'University Library Information Services and Resources for Engineering Distance Learners'. The 'Information for Staff' area is only for authorised Loughborough University staff. Students have no access to read documents in this area. Module Leaders post module outline plans for their module in this area prior to the residential week. 'Questions and Answers' is an area where students can ask any general or technical questions which are not directly module related.

The 'Chat for Students' area is a discussion area for students. The lecturing staff have no access to write or read the documents in this area. Students can create their own documents in this area. Once a student has created a document, his or her user name will appear with the title of the document. The aim of this area is to provide a communication facility between fellow students on the course. However, in comparison to the module related discussion areas this area is not extensively used. A possible reason for this is that the current students are all employees of the Ford Motor Company, Jaguar and Aston Martin and therefore communicate at work, either face to face or by other means.

The most active areas of the CBSSS are the 'Study Support Materials' and 'Discussion Forums' areas. Each of these areas have been divided into three categories as seen in Figure 2. Within each category there are sub-categories under the relevant module names. The 'Study Support Materials' areas are used to deliver the pre- and post-residential study support material. Most of documents are distributed as attachments within Lotus Notes documents. The majority of the attachments are in Adobe Acrobat PDF format. The other attachments include Excel data sheets, text files and computer programs written in C, FORTRAN, Matlab and Simulink. Since attachments preserve the formatting of the documents, students can use the data files and computer source codes without having to reformat them. Lecturing staff involved with the MSc course can create documents in the CBSSS with a very little experience in using Lotus Notes.

The 'Discussion Forums' are the most widely used areas in the CBSSS. There are dedicated discussion areas for individual modules. Students and lecturing staff can write in these areas using the edit facility provided within the system. However, since the World Wide Web is being used, there are some limitations. For example, currently, users cannot enter mathematical equations using an Equation Editor or type mathematical expressions with symbols, superscripts and subscripts. Staff and students must express mathematical equations using normal text. However, neither students nor staff criticised this as a problem or a drawback. With future software upgrades this drawback may be overcome. If lecturing staff need to add material with equations and diagrams into the discussion areas, they can use Lotus Notes to edit the relevant discussion area. If necessary, lecturers add explanatory material into the 'Study Support Materials' area and give the document location details in the appropriate discussion area. With the CBSSS, staff can provide effective study support to students at the work place, by communicating with them through the discussion areas and by putting additional study support material onto the system.

The discussion areas are common areas for all students in any particular Team. All the students in that Team can read the entries in the discussion forums. A group of six students who have been using the discussion areas for more than a year were asked if they would use these areas more if the discussion areas were private. They all said it would make no difference to them. They were also asked if there are any features in the CBSSS which could be removed. The overwhelming response was that all features should be retained.

The main drawback of the system is that the WWW can be slow at certain times of the day. To overcome this a new infrastructure was recently introduced which includes the hyperlink 'What's New/Modified'. This eliminates users having to look through all the areas in the CBSSS to find new or modified documents and thus speeds access time. The CBSSS was introduced at the end of 1997 and is still being developed. It is also anticipated that more facilities will be introduced in the future.

5   DISCUSSION

The MSc in Automotive Systems Engineering has now replaced the previously taught MSc in Advanced Automotive Engineering, the aim being to provide a top-down structure for the automotive engineering content, based on such things as system level requirements. Planning these changes has been a joint project between The Ford Motor Company and Loughborough University. Vehicle engineering and manufacturing content both form part of this wider picture, though the emphasis of the programme remains on the vehicle side.

Two years ago there began further discussions with industry representatives aimed at cutting the cost of the course, whilst preserving academic quality and content. One of the main concerns highlighted by the discussions, was the very real cost of engineers spending significant periods of time away from their company. Hence, the decision was taken to reduce the residential time at Loughborough. A number of models were discussed, but it was felt by most people that delivering entire modules by Distance Learning was not desirable. One of the many reasons was that the 'networking' achieved among students via the residential modules was exceptionally valuable, and that the sense of isolation that would be experienced by individuals studying at a distance should be avoided. Hence a plan was introduced which reduced the residential periods to one-week, and supported company based study with an on-line delivery system. This modern flexible/distance learning system is based on Lotus Notes and the Internet. It is currently being used in the programme to assist student communication with the University and it is being further developed and improved for the future.

Initially there were a number of reservations expressed by staff and students about the new delivery format. The following questions highlight the main concerns expressed about the new programme and format:
- will the depth and quality of the course suffer?
- will lectures be hurried or compressed?
- will students have to study 'distance learning' packages?

First and foremost, Loughborough University will continue to award an MSc degree at the end of the programme, and there is no freedom to downgrade this award, so the depth and quality *cannot* be reduced. The new delivery mechanism ensures that the depth and quality of the course does not suffer. Secondly, the lectures should not be hurried or compressed because the 'equivalent' lecture time is the same, merely the delivery format has changed. Finally, self contained distance learning packages are not used to any great extent. This is a masters level programme and a reasonably high level of maturity is expected of the students. Thus, the use of step-by-step programmed learning methods is considered generally inappropriate. However, some of the early study material in the core modules is covered in this more 'traditional' distance learning approach.

The future structure and delivery of the programme is under a system of continuous improvement. This is embodied in the management of the MSc programme. Of particular importance are regular meetings with an industrial liaison committee - a team of industrial managers and technical experts who works with staff from Loughborough University. This committee helps to steer the MSc in the design, development and management of the programme.

6 ACKNOWLEDGEMENT

The assistance of industrial managers and technical experts from The Ford Motor Company who have given many hours of their valuable time to help in the design and development of the programme is gratefully acknowledged.

Table 1 Modular structure of the MSc programme

	Assessment		MSc Credit
	Exam	Coursework	
Core Modules			
Engineering Framework	50%	50%	15
Vehicle Systems	50%	50%	15
Systems Engineering	50%	50%	15
Manufacturing Systems	50%	50%	15
Elective Modules *(Select Four)*			
Vehicle Dynamics	- - -	100%	15
Manufacturing & Material Processing	- - -	100%	15
Vehicle NVH	- - -	100%	15
Powertrain Engineering	- - -	100%	15
Vehicle Platform Engineering	- - -	100%	15
Engine Performance & Design	- - -	100%	15
Design Integrated Manufacture	- - -	100%	15
Automotive Control	- - -	100%	15
Advanced Automotive Materials	- - -	100%	15
MSc Project	- - -	100%	60

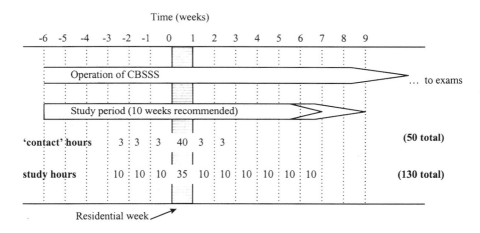

Figure 1 Module delivery format

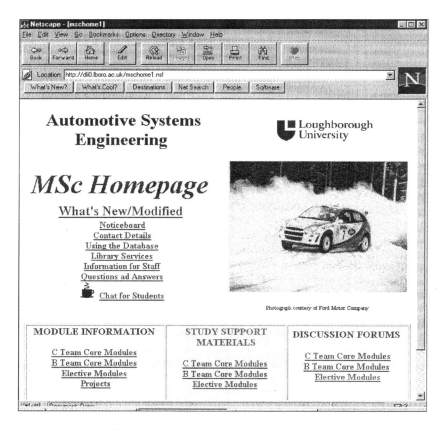

Figure 2 Screen dump of the MSc homepage

C574/021/99

The virtual automotive learning environment

P R BULLEN, F HADDLETON, P B TAYLOR, and **M YOUNG**
Department of Aerospace, Civil, and Mechanical Engineering, University of Hertfordshire, Hatfield, UK
P F CHATTERTON
Daedalus, London, UK

ABSTRACT

The development of multimedia Information and Communications Technology (ICT) applications to automotive engineering education programmes is described providing details of the structure and outline content of these applications. They have been developed for a Masters programme for employees of the industry and have now been extended to a pilot project involving automotive supply SMEs. The effectiveness of these applications has been evaluated demonstrating benefits in learning and a reduction in time away from the workplace for learners. Resource implications and the need for collaboration between education providers and the industry have been identified.

1. INTRODUCTION

The University of Hertfordshire runs a Masters Degree programme in Automotive Engineering Design, Manufacture and Management (an integrated graduate development scheme, IGDS) for employees in the automotive industry. This programme is organised on a modular basis designed for part-time study. The programme consists of a number of courses, most courses being arranged as a five-week block containing a residential component of up to five days, depending on the course. The programme is very successful (1) but it was recognised that there were a number of issues that needed to be addressed to meet the needs of our customers. These were:

- To reduce time away from the work place required for short course attendance
- The need for increased flexibility in study patterns
- To encourage delegate interaction outside the residential course
- To increase the opportunity for tutor delegate interaction outside the residential course
- To decrease the cost of the programme by reducing the costs of residential accommodation

To meet these needs a project was set up, supported by the Ford Motor Co., to introduce technology – enhanced distance learning and delegate support - into the programme. In addition the five partner Universities involved in delivering the programme recognised potential benefits in introducing multimedia Information and Communications Technologies (ICT) and formulated the following educational objectives, as well as highlighting benefits in terms of administrative support and inter-university communications.

- To expose delegates to creative synthesising and open-ended problem solving experiences.
- To emphasise processes and facilitate group project work and brain-storming.
- To develop high quality educational materials
- To raise awareness of external information sources such as electronic databases, libraries, document delivery services and provide desktop access to these sources.
- To provide 'on-demand' student support, help and guidance.

In parallel with the above developments a group known as VACU (Virtual Automotive College and University) was being formed. This group comprises representatives from HE (including the University of Hertfordshire) and FE institutions, from the DTI Automotive Directorate and from the SMMT Industry Forum, which is focussed on sustainable education and training applied to SMEs (2). The potential benefits and difficulties of applying Information and Communications technologies (ICT) within the SMEs and this group was established during a feasibility study (3) undertaken in April '98 to assess the needs for ICT solutions to support the activities of VACU. This study led to the implementation of an online support system, developed from the experience gained with IGDS, for an ESF, Objective 4 supported pilot project to deliver project management and SPC sustainable learning to eight SMEs in the West Midlands region.

2. DEVELOPMENTS TO SUPPORT THE MSc PROGRAMME

The following developments in multimedia ICT applications have taken place in support of the educational objectives of the Masters programme and the business objectives of Ford and of the Department of Aerospace, Civil and Mechanical Engineering at the University.

2.1 Development of multimedia CD-ROM materials
The specific objective was the restructuring and conversion of an existing, five-day residential course (Automobile Materials, Manufacture and Processes) into a computer-based, interactive distance learning package for delivery on CD-ROM (4). A number of automotive manufacturing and supply companies agreed to become involved in the project and have given permission to access and use company information (library facilities, product information, data, diagrams, photographs, videotapes etc.). To date 60% of the course has been converted to multimedia format.

2.2 Development of online communications and information-sharing infrastructure

An online communications and information-sharing infrastructure has been developed based around the groupware package, Lotus Notes (Domino).

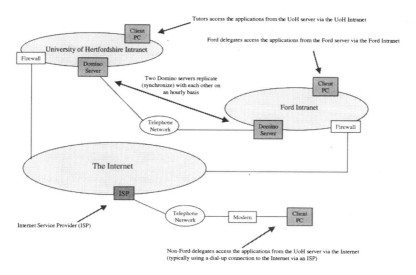

Figure 1 Schematic of Intranet and Internet Network

This infrastructure supports the development of learning and administration/management applications which are based on Lotus Notes databases that can be accessed either via Lotus Notes client or a standard Web browser, illustrated in Fig. 1.

2.3 Development of online learning applications for courses.

2.3.1 Programme Forum

A main IGDS Programme Forum has been developed to enable all tutors and delegates on the IGDS programme to communicate freely with each other using any Web browser. One of the major benefits of the Forum over more traditional means of communication is that it promotes group/collaborative discussions and support and, at the same time, provides a permanent record of the discussions. The course Forum features the sections shown in Table 1.

Table 1 Programme Forum

News	The News section is for everyone to contribute any interesting news items that might be of interest to others – e.g. an upcoming TV program or forthcoming events in the automotive industry
Introduction	A guide to using the system together with general information about the course.
Participants	This is a list of participants on the MSc program – which is completed by the students, tutors and administrators. It gives contact details, a photo and a section for hobbies and interests. In addition, participants complete a survey of their IT systems at work and at home – enabling a comprehensive database to be constructed of delivery IT systems.
Materials	This section is under construction and will be used by tutors to publish course materials or course notes.
Discussions	The discussion section is used to discuss issues, pose questions and generally communicate in the form of a bulletin board or newsgroup.
Library	The library is used as a repository of useful references for everyone to contribute to, such as books, Web pages, or articles. There's also a section for your comments, reviews or abstracts. When students submit assignments, they are asked to submit their references in a standard form whereby the data can be added to the library. Over time, it is anticipated that the library will build up to be a valuable knowledge base on the automotive sector.
FAQs	This section provides answers to "Frequently Asked Questions" (FAQs) – either on academic topics or on technical issues.
Glossary	This section will build up to form a relevant glossary of terms.
Evaluation	Students can complete an evaluation form to make general comments on the course and the online applications. Each module has its own tailored evaluation form.
Menu	Menu of links (hypertext) to each individual Course Forum.

2.3.2 Individual Course Forum
From the main Programme Forum menu, delegates, registered on specific courses, are able to gain access to individual Course Forums, enabling these students to access course specific information. In addition, this Forum is used by delegates to submit assignments. Tutors use the Forum to supply supporting material, interact with delegates on subject specific issues and provide feedback on assignments. External Examiners are also given access to the system so that they have an oversight of the programme, course materials, assessment process and of the quality of delegate work.

Each Course Forum also has an evaluation form where delegates are encouraged to provide feedback on the course. This information is then used to inform the annual programme monitoring and evaluation, including the course review process. The Forum provides a permanent record of assignment submission and delegate comment and provides a convenient way of capturing feedback and discussion on subject issues.

2.4 Development of online administration and management applications and desk-top video-conferencing

A number of databases have been established to support the management and administration of programmes and courses. These include, document management, meetings management, task management, delegates' records, multimedia resources, a 'green book' of best practice and discussions sections.

Three desktop video-conferencing systems have been installed. These are PictueTel systems that operate over the public ISDN network and are similar to those used extensively throughout Ford. The systems have been used to aid the development of some of the applications described above and to support course development.

These developments have resulted in CD ROM based material for one option course and a system supporting the whole programme known as VALE (Virtual Automotive learning Environment) whose effectiveness is evaluated below.

3. AN EVALUATION OF VALE

3.1 Multimedia materials

Overall the delegates on the Automobile Materials, Manufacture and Processes course found that the multimedia materials were of high quality and more effective than conventionally taught materials. The key feature was the interactive nature and the capability to study at their own pace, and in their own time, either at work or at home, selecting appropriate parts of the materials for their needs. They also commented on the ability to access a large amount of subject matter very easily. However the numbers of delegates choosing this option course has declined since it has been in this multimedia format indicating a general reluctance of delegates to choose study in 'isolation' if they have a choice between distance learning and a residential short course. This identifies a possible conflict between the needs of the employer and the employee.

One of the major lessons learnt is that the development of multimedia materials is a very time consuming and resource intensive process. Typically the time requirements are of the order of 100 hours for every hour of conventional delivery. There are benefits in the quality of learning and the saving in the reduction in delegate time away from the workplace but the economics make it difficult for an individual tutor to develop multimedia materials. There is a need to see these developments within a larger scale project involving a number of institutions with potential benefits to a number of companies.

3.2 Online learning applications

The main programme Forum was introduced in October 1997 and individual course Forums introduced as each course commenced. This is still an ongoing process. Ford delegates access the Forums through the Ford Intranet, whilst non-Ford delegates use the Internet. The Forum used in the delegate's first course, on the programme, was more structured using discrete areas with greater guidance given to delegates on how to use the Forum. Delegates were also required to submit interim assignments to establish tutor delegate discussion at an early stage in the programme and to 'force' use of the system.

An evaluation of this learning application was undertaken by an open (face-to-face) discussion and by using online evaluation forms. The feedback showed that the majority of delegates found the system useful and particularly appreciated the delegate-delegate and delegate-tutor interaction outside of the residential component of the course. Delegates were unanimous that the system should be continued and further developed. A system has now been instigated into the monitoring and review process of each course whereby tutors are asked to demonstrate how they will increase the use of the system when their course next runs in order to fulfil the specific educational and business objectives of the course and programme.

3.3 Video-conferencing

Use of video-conferencing was restricted to audio-visual communications in the early phase of the project but more recently has included document sharing. The technology supports greater dialogue 'at a distance' and saves time lost to travel. The next step is to introduce the use of video conferencing for more formal meetings, for example for subject boards of examiners involving partner Universities and for tutor delegate discussion. A pilot project was carried out involving delegates in the UK and in the USA, facilitated by tutors, involving a case study on cultural issues in a multi-national engineering team. This demonstrated the feasibility of running a team project with team members based in different locations but also showed the need for good preparation for, and good organisation of, the video conference.

3.4 Conclusions

The experience gained to date has clearly demonstrated the benefits of applying Information and Communications Technologies in support of educational programmes in particular where the delegates are largely based in different companies, or in the same company but at different locations, distant from the tutors. The benefits of using such systems are further realised when such programmes are delivered by a consortium of Universities, for example. However these developments, in particular the development of multimedia material, are time consuming and resource intensive. There is a need to consider carefully the learning model and objectives of the programme and individual courses within the programme, and the needs of the customer, and to use the technology appropriately to both enhance learning and reduce overall costs.

4. APPLICATION OF INFORMATION AND COMMUNICATION TECHNOLOGIES (ICT'S) TO EDUCATION AND TRAINING IN THE AUTOMOTIVE SUPPLY SECTOR

A feasibility study was commissioned by the DTI (3) to assess the needs for ICT solutions to support education and training in SMEs. A key element of the study was a survey of 23 companies and 7 course providers using pre-prepared questionnaires to conduct personal and telephone interviews. In some cases course providers chose to complete an online questionnaire.

The survey revealed that there are strong commercial and operational drivers for SME companies and colleges/universities to implement flexible learning techniques - most notably to increase the quality and efficiency of learning and to minimise the time that SME employees are taken out of operational roles whilst undertaking courses. These are similar drivers to those identified in 2. above for employees on Masters courses.

However as has already been indicated, the resources required to rapidly implement flexible learning techniques on a significant scale are beyond the capabilities/resources of individual educational institutions. It is therefore concluded that an industry approach could accelerate the implementation of flexible learning techniques particularly within the SME community involving collaborative developments, sharing of resources and expertise and implementation of best practice processes.

The survey also highlighted awareness issues amongst both SME companies and colleges/universities:
- SME companies; a significant number lacked awareness of the needs for and value of learning programmes and techniques that can help them to update working practices to meet the quality and productivity-driven requirements of the vehicle manufacturers.
- Colleges and Universities; a general lack of awareness of how to cost-effectively exploit ICT systems in developing quality flexible/distance learning programmes.

The penetration of IT systems amongst SME companies was found to be higher than anticipated at the outset of the study - with over 60% of companies interviewed providing nearly all managers with access to a PC. However, there is considerably less access for non-management staff. On average for all companies, 50% of PC's featured a multimedia capability, however Internet access is typically restricted to one or two PC's mainly used by managers.

Attitudes towards using technology-based learning were generally favourable with many SME companies having some (limited) experience of computer based training programmes. There was a general view that such programmes would be welcome, provided the content and price were appropriate.

It was recommended that a virtual Intranet (VACU Online) for the automotive industry should be developed for use by SME companies, vehicle manufacturers, course providers and other industry bodies. Figure 2 below illustrates some of the key features of the recommendations.

The purpose of VACU Online is to:

- Facilitate sharing of resources, expertise and knowledge for course development, delivery and administration.
- Provide VACU standardised tools and resources.
- Promote faster communications and problem solving capabilities.
- Help disseminate information and awareness of issues
- Co-ordinate best practice and quality processes.
- Promote the industry as offering 'high tech' learning.

It was also recommended that the first step in such a development should be in support of the Virtual Automotive College and University (VACU) pilot project.

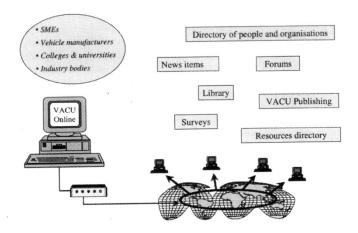

Figure 2 VACU Online Features

5. THE VACU PROJECT

The thinking behind the "Virtual Automotive College and University" (VACU) goes back to a 1995 report on the competitiveness of the UK automotive supply chain. This concluded that UK firms were being put at a competitive disadvantage by a shortage of suitably qualified engineers. A "Virtual Automotive University & College" was suggested as a possible solution.

The initiative has been taken forward by a Steering Group including representatives of universities, further education colleges, automotive firms, the Society of Motor Manufacturers and Traders Industry Forum (SMMT IF) and the DTI. Its aim is to upskill key groups of employees within the UK automotive components sector through a collaborative approach to developing and delivering training and development programmes – with a top-level aim of improving performance in QCD (Quality, Cost and Delivery) - the essential measures of competitiveness.

To-date, VACU has focused on the development and delivery of training programmes in Project Management and Statistical Process Control (SPC). These programmes are customised to suit the needs of each SME through an initial "diagnostics" process and then delivered in a flexible methodology that overcomes some of the practical constraints that SMEs face in providing sufficient time and resource for their employees to undertake training. Throughout the delivery, constant collaboration and feedback is encouraged.

It was recognised, in view of the multi-partner nature of VACU, that ICT applications would be required to support its activities. A strategy was therefore adopted that ICT applications would initially be focused on supporting the administration, management and course development functions of the VACU steering group and academic team members. The development of these applications was based on experience gained in developing the VALE system described in 2. above.

The first application developed was an Online Information Management (OIM) system, the purpose of which is to act as the primary means of electronic communications and information management for the geographically dispersed team members (located throughout the UK). The OIM system is accessed via the World Wide Web and is password controlled in order to restrict access to team members only - and is divided into the following areas:

General Discussions area
Team members hold general discussions in this area – working in the same way as Internet newsgroups or bulletin boards.

Meetings area
Meeting agendas are posted in this area and meeting minutes are also recorded together with comments on the minutes.

Tasks area
The Tasks area is used by the Project Manager to detail and monitor all tasks. Tasks that are overdue are highlighted.

SMEs and beneficiaries area
This area records details of the SMEs and the SME beneficiaries (trainees) that participate in the training programmes.

Documents area
All project documents are stored in this area providing a single and manageable repository of project documentation.

Library area
The Library area is a database of useful references and links to sources of information, such as books, journals, news articles and Web sites. All team members are encouraged to contribute towards the Library with the intention of building up a valuable knowledge base that can be exploited in learning programmes.

Feedback
A feedback area is provided with a number of online forms that team members can complete, for instance, to provide feedback on the features and effectiveness of the OIM system.

Planning area
This area is similar to the discussions area, but the discussions are focused on planning issues with respect to the development of the Automotive College.

Materials development area
This area is used to manage the development/production of course materials, such as course notes or open/distance learning materials. Developers can upload drafts of course materials for comments and review by other team members. The OIM system automatically version controls drafts and users can choose to review either the latest versions of materials or to inspect an audit trail of the development cycle.

The OIM system has been highly successful in maintaining a high degree of collaboration and communication between the geographically dispersed multi-partner group and will become increasingly important as the membership of the group grows.

In early 1999, VACU commenced a project to develop quality learning materials. These will be in the form of a tutor pack where tutors can "pick & mix" component materials to create a student pack that is customised to each SME. ICT will be used extensively both to manage the development/production of the learning materials and as a delivery medium. For instance, the OIM system will be used to manage the development of the materials, which, when complete will be stored onto CD-ROM. Tutors will have the option of creating hard-copy materials or, for example, to run presentation slides via overhead projectors. Trainees will be able to access the materials on the Web though this is not seen as the primary means of learning materials delivery.

The next phase of ICT applications development will be to open up access for SMEs and beneficiaries to the online applications – in the form of a "Support Forum". This will feature similar areas to the OIM system, such as Discussions, Library and Feedback. It is intended that SMEs and beneficiaries will be able to access this after the cessation of the formal training programmes – allowing tutors to offer ongoing trainee mentoring as well as allowing beneficiaries within different SMEs to develop and maintain contact with each other.

The long-term vision of the ICT applications is to develop an online automotive community where SMEs, vehicle manufacturers, the SMMT Industry Forum, Universities and Colleges and government are actively collaborating together – exchanging and sharing knowledge and working and learning together.

6. CONCLUSIONS

- Information and Communications Technologies (ICT) applications and multimedia materials have been developed in support of educational programmes for the automotive industry.
- An evaluation of these applications, developed to support a Masters programme has demonstrated the benefits in terms of enhanced learning and improved delegate support.
- The experience gained with the Masters programme has led to applications of the system to support partner organisations in the Virtual Automotive College and University in a pilot project to deliver education and training to SME automotive supply companies.
- Issues have been raised with regard to the resource requirements of these developments, the need to form collaborative partnerships and the need to use the technology to meet educational and business objectives.

REFERENCES

1. Bullen, P., Mughal, H., and Taylor, P., 'Developing Engineers in the Automotive Industry', International Conference on Education in Automotive Engineering - Autotech 99 – NEC, November 1999.

2. Barlow, N., Lyons, A., Jones, M., Davey, R., Chatterton, P. 'Sustainable Learning in the Automotive Supply Chain', International Conference on Education in Automotive Engineering - Autotech 99 – NEC, November 1999.

3. Bullen, P., Chatterton, P., 'Development of IT Systems to support the activities of the Virtual Automotive College and University (VACU)', Report to the DTI, April 1998.

4. Haddleton., F., Taylor, P., and Chatterton., P., 'Course Design and Delivery Using New Technologies and an Evaluation of Educational Effectiveness', The Proceedings of the 1997 ASME Mechanical Engineering Heads Conference, March 1997.

ACKNOWLEDGEMENT

The support and help of Professor G. Johnson, University Programmes, Ford Motor Co. is gratefully acknowledged.

Skills and Competencies

C574/011/99

Critical competencies for automotive engineers – how to identify and develop the necessary skills

P B TAYLOR and **P R BULLEN**
Department of Aerospace, Civil, and Mechanical Engineering, University of Hertfordshire, Hatfield, UK
J A MULRYAN
Rover Group, Lighthorne, UK

ABSTRACT

This paper outlines the procedures undertaken by Rover Group to establish the skills required by Engineers throughout the Design and Engineering function. It will then go on to discuss the some of the key findings of this exercise and some of the limitations of the data collected. The outcome of this process was a framework called the 'Critical Competence Standards Framework' that is now being used to design and develop an educational scheme to cover 'gaps' in an individuals skill base thereby giving the engineers the skills they need to carry out their job function.

1. BACKGROUND

1.1 Rover Organisation

The Rover Group Design and Engineering function is an organisation of approximately two and a half thousand engineers responsible for the design, test and development of new and existing vehicles, systems and components within the Mini, MG, Rover and Land Rover marques. It is comprised of four 'Centres of Competence' (CoCs) namely:

- Chassis
- Electrical and Electronics
- Body and Trim
- Vehicle Integration

Within this function there are also Vehicle Project Teams who are responsible for management of new product programmes. It should be mentioned that at the beginning of this project Power Train Design was a separate business unit within the Rover organisation and so is not included in this exercise.

It was clear from the outset that to establish the professional competencies required across the organisation would be a significant piece of work, due to the diversity of both the job responsibilities across the CoCs and of individual responsibilities within each part of the organisation. A further complication was the educational background of the engineers, which varied greatly from City and Guilds, ONC, HNC and degrees to MBAs and Doctorates.

1.2 "World Class Skills" Initiative

The project formally began in September 1996, when it was decided that action needed to be taken to address the perceived skills gap within Design and Engineering. This skills gap was apparent when compared to equivalent graded colleagues at BMW and against the skills required to meet company objectives in terms of technical content and quality of current and future vehicles. The reasons for this were felt to be that these skills had become diluted through years of working in partnership with Honda who had retained technical responsibility on all collaborative projects. Additionally Rover had become more dependant on the technical capability of engineers within the component supply industry. Whilst the company recruited and trained able graduates and apprentices, basic skills and capabilities were lost through lack of use and application in a real life situation. Rover have built up an excellent reputation for encouraging individuals to continue learning through Rover Learning Business and using a wide variety of external organisations, but there was little training available through the company to address the technical skills required by engineers in today's increasingly competitive environment. There was a recognised need and desire to go back to basic scientific and engineering principles in the design and development process, and to encourage innovation and creativity through application of these first principles.

Under the leadership of staff from the Learning and Development Department an initiative called 'World Class Skills' was launched to identify the skills required within the Design and Engineering Function. This initiative was in line with the Rover Company quality strategy QS2000 that states that 'all training will be competency based'. The philosophy behind this statement originates from the Experiential Learning Model (1), which provides a framework for examining and strengthening the critical linkages between education, work and personal development. This model uses a system of competencies for describing job demands and corresponding educational objectives.

The first task in this project was to develop a standards framework for each CoC. World Class Skills were to be developed in two stages:

- 'Core' competencies - those skills which it was felt could be developed to build upon the successes of the Rover Group marque values, for instance, further refinement in the vehicle ride and handing area;

- 'Critical' competencies - those skills that would incorporate fundamental engineering skills and knowledge and thereby produce vehicles that meet customer expectations in terms of quality, performance and cost.

Whilst recognising that management and behavioural skills are important in the design process these were deliberately omitted from this project as these skills are addressed through other training programmes already available within Rover Group.

The work described here concentrates on the development of the critical competence standards framework although a similar approach was used for both schemes.

Firstly existing standards were reviewed. Similar standards currently available in the engineering and manufacturing field include 'The New Engineering Occupational Standards' (1998) and 'Engineering and Marine Training Authority (EMTA) standards'. National Vocational Qualifications (NVQs) were designed to represent 'a statement of competency confirming that an individual can perform to a specified standard in a range of work-related activities, and also possesses the skills, knowledge and understanding which makes possible such performance at the workplace' (2). Consequently NVQs were examined in some detail, as the aims were similar to those required by Rover.

Unfortunately the content and detail within all these existing standards was not considered suitable or sufficiently objective to be used as the basis for measuring the competencies as required by Rover or for future development of training provision. There are currently no specific automotive standards available.

2. DEFINING THE SKILLS NEEDED

In 1994 Rover Group had successfully launched a Modern Apprenticeship scheme which incorporated General National Vocational Qualifications (GNVQ) at Foundation, Intermediate and Advanced levels. This included NVQs at levels 2, 3 and 4. This experience and the common aims discussed above, led to the adoption of an NVQ-style approach with each CoC developing a standards framework for their areas of specialisation. With the assistance of EMTA the following process was defined to firstly develop functional maps and then to write in the elements of competence.

2.1 Develop a Functional map
Develop a key purpose statement which defines the fundamental role for the CoC. For example in the case of the Chassis Department this was 'to deliver fully proven Chassis systems and components that meet product targets.'

2.2 Define Units of Competence
This split the key purpose statement into the major tasks or Units of Competence, to be completed by the individual CoC in support of the design and development process of a new vehicle from concept through to volume production.

2.3 Define Elements of Competence
Elements of Competence were then developed which define in greater detail the tasks to be completed within each Unit of Competence. Rover Group personnel were advised by EMTA that each unit should be broken down into approximately five elements which meant that each had sufficient content to stand alone and yet still define a substantial task. Five elements also meant the each would not become too large and therefore over complicated. From this point, it should be possible to use these units and elements to define a particular job function. All

engineering and technical activities taking place within the organisation were considered, including research and pre-development.

Figure 1 shows an extract from the standards framework for the Electrical and Electronic CoC which shows the tree structure used to link the Key Purpose, Unit of Competence and Element of Competence. The individual elements of competence were written after careful consideration and analysis of the tasks that are carried out at each phase of a vehicle development programme. Detailed knowledge of Rover processes and culture were used to ensure that this first draft document was as accurate as possible.
Each element was then developed into an NVQ-style format as below:

2.3.1 *Performance Criteria*
Developed as statements of competence which an individual engineer would be expected to achieve. Again for simplicity and completeness, EMTA suggested that each element should contain six to eight performance criteria statements.

2.3.2 *Range*
Range statements were used to clarify terms used in the performance criteria - this was particularly important when referring to Rover processes or using Rover terminology. They also allowed for greater detail to be given which could not be included in the performance criteria.

2.3.3 *Underpinning knowledge*
This section defines the skills and knowledge required in support of performance criteria and has been taken further to develop the necessary training and education programmes.

2.3.4 *Supportive learning material*
A list of training sources which was then cross referenced to the underpinning knowledge and could be used by Managers to identify suitable training for a team or an individual.

Figure 2 shows how the Element of Competence: 'Assess feasibility of alternative design solutions' from Figure 1 is expanded. The Performance criteria, Range and Underpinning knowledge are shown, illustrating the relationship between the different levels of information.

The standards frameworks are currently being modified to further define the underpinning knowledge as distinct knowledge and skills. This is useful in understanding the subtle difference between knowing a theory and being able to apply and use it. Knowledge and skills will be cross-referenced to Performance Criteria and supportive training material. It is hoped that this will prompt greater thought by a Manager or individual when identifying skills gaps and training needs and will focus on the business needs.

Each CoC nominated individual Senior Engineering Managers who would work with Managers from the Learning and Development Department to develop the standards framework for their functional area. The task was greatly simplified by the decision to write a common first draft framework for Chassis, Electrical and Electronics and Body and Trim. This was then modified to incorporate specialist skills and knowledge or reflect varying work practices. This proved to be extremely successful, and to many people's surprise, identified many areas of commonality across the different CoC's. Modifying this common framework and including specialist skills as required, led to final versions for each CoC.

3. KEY FINDINGS

3.1 Specification

The process of writing standards frameworks was extremely thought provoking and beneficial. It may at first have appeared simpler to ask representatives from the CoCs what their training needs were. However whilst generating a list of courses required, this would not have produced an understanding of why these knowledge and skills were needed or how they would be used in the workplace. Generating the Units of Competence was also educational in itself, revealing a wider range of skills than was originally thought and significant areas of commonality.

The first published version of the standards framework has been extremely well received by EMTA who have expressed an interest in using this information as a starting point for the generation of automotive standards. They were also issued to the University of Hertfordshire for the design and development of a new training scheme for engineers within Design and Engineering. Responses to this reflected a clear understanding of the objectives of the training scheme and of the Rover organisation. Subsequent development of the scheme with the University of Hertfordshire also reflects the accuracy of the original document

3.2 Skills identified

There was a high level of commonality in underpinning knowledge identified across the standards frameworks. The use of Learning and Development staff, who already had knowledge of the organisation and its processes, to facilitate the process enabled a pragmatic and objective approach to be taken. The concept of writing competence standards was in some areas seen as unwieldy and in others excessive detail was included. An essential element to the success of this work was the enthusiasm and ability of the individuals involved.

The standards frameworks must be viewed as a live document which can be regularly reviewed and updated to reflect the changing skills requirement presented by changes in technology and market requirements.

3.3 Limitations

Whilst common skills were identified, the level and type of knowledge or skill were not necessarily clearly defined. For example skill in electrical engineering was highlighted in a number of areas but the skill level required in each CoC varied. An understanding of basic laws and principles which could be applied to design and analysis of simple circuits is required by Chassis and Vehicle Integration, whilst in the Electrical and Electronics function a higher skill level is required, including circuit and software design.

An additional reason for using an NVQ format for standards framework was that it may be decided at a future point to gain accreditation for the units and thereby enable engineers to gain additional qualifications with little additional effort. Documentary and practical evidence from an individual's work for the company could be submitted as part of their portfolio. Whilst this may still be a long term objective there is a practical difficulty of finding assessors of suitable competence in the wide range of skills identified.

4. DEVELOPING AN EDUCATION PROGRAMME TO MEET ROVER'S NEEDS

4.1 Academic Background

Many of the competencies identified by Rover were already part of a part-time, modular MSc scheme which had been developed specifically for, and in partnership with, the automotive industry in 1994. This scheme is a MSc. in Automotive Engineering Design, Manufacture and Management (3) which can be taken as individual short courses for continuing professional development, or over time, built up to a complete Masters Degree. Additionally this scheme is delivered by a consortium of Universities, led by the University of Hertfordshire, building on the individual subject strengths of each of the partners: The Open University, Birmingham University, University of Luton and Loughborough University. This previous experience led to Rover choosing the University of Hertfordshire and this consortium as their educational partner for the 'Critical Competence' programme.

The first activities undertaken were to identify the entry and exit points required by Rover, that is to say what range of qualifications/background/experience etc. the engineers entering the scheme would have and what was required to be added to this through attending the scheme. It soon became apparent that there would be a very wide range of entry levels and there must also be a wide range of exit points ranging from individual certificates, to recognised academic qualifications such as Post-graduate Diploma and Masters Degree. There also needed to be a consistent way of measuring the skills/knowledge added through the scheme so as to prove that the required competencies had been gained and were being implemented 'on the job'.

4.2 Key Features

The design of this scheme borrowed many features from the existing MSc which was 'tried and tested' and met many of the criteria required by Rover. Consequently the Critical Competence Scheme was designed to be:

- Modular:
  with each subject module being a 'stand alone' course with it's own assessment built in.
- Flexible:
  to fit around an individual's needs and be compatible with job requirements.
- Work-based:
  each individual would bring issues to the course from their job which would be examined as part of the course work, and problems would be solved as part of the assessment.
- An academic-industry partnership:
  each module was designed and delivered by a team comprising academic staff from the University involved, and relevant subject specialists within Rover.
- Good quality:

using closed-loop feedback procedures to ensure the quality of the educational provision. Established continuous improvement processes were adopted and joint scheme management was undertaken. (4)
- Industrially relevant:
all delegates would be expected to use the learning from the course in their day-to-day work and be able to show where this had happened.

4.3 Work to Date

To date the structure of the scheme has been established, some twenty module titles have been agreed and joint working groups have been set up to design and deliver each module. The Management and quality assurance structures have been agreed and delivery of the first modules is expected to take place shortly. Rover are currently working on identifying the most important subject modules and individuals who need to take them as a priority to meet the needs of the business. This process will be linked to the established appraisal procedure within Rover where an individuals developmental needs are established. This strategy is recommended as the way to integrate appraisal as part of wider employee resourcing and development strategies and policies (5) and again becomes a closed-loop system.

5. CONCLUSIONS

This method of working, while very detailed and time consuming, has produced an excellent framework and set of competencies, which can be applied throughout the automotive industry. Each Department within a company can adapt the competencies identified by Rover for their own needs, and then take this forward to identify the educational route which will provide specific individuals with the skills/knowledge required to enable them to carry out their job function.

Partnerships with appropriate specialist institutions, such as the University of Hertfordshire, can then turn these requirements into industrially relevant educational programmes and contribute to the 'bottom line' performance of the company. The benefits to the company are found to be wider than just up-skilling staff and include networking across the industry, sharing 'best-practice', learning from others practical experience, increasing employee motivation and interest, re-awakening of educational ambitions and lead to the goal of lifelong learning.

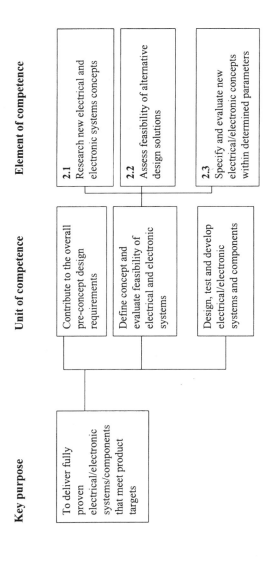

Figure 1 Sample of Standards Framework for Electric/Electronic CoC

Ref EE2.2   Assess feasibility of alternative design solutions

PERFORMANCE CRITERIA (defining the standards to be achieved for satisfactory performance of the element of competence)

1. Alternative design solutions are evaluated against **functional specification** for the component or system

Strengths and weaknesses of alternative design solutions are identified:
1. Alternative design solutions are assessed for **serviceability**
2. Alternative design solutions are assessed for <u>feasibility</u> of manufacture & assembly to required tolerances by the <u>supplier</u>
3. Performance of alternative design solutions against **design objectives** is established

RANGE (describing the circumstances, conditions and applications in which competence must be demonstrated)

Functional specification: that which describes the operation of a component or system
Feasibility: considers manufacturing process capability, ability to deliver
Supplier: Rover group internal or external sources

UNDERPINNING KNOWLEDGE (identifying the requirement for knowledge of relevant principles and methods of appropriate data and information necessary for competent performance)

Knowledge	Skill
Design processes	Design for assembly
Manufacturing processes; metal, plastic, electronic	Design for manufacture
Assembly processes; metal, plastic, electronic	
Tool design	Application of electronic theory & principles
Electrical theory & principles	Application of electrical theory & principles
Analogue & digital electronics	
Use of test procedures within a manufacturing environment	
Optical networks	
Evaluation methods	

Figure 2 CoC Element of Competence

REFERENCES

1. Kolb, D., 1984 'Experiential Learning' pub. Prentice-Hall, New Jersey.

2. Department of Employment. Employment News, No. 61, June 1988 'Education and Training for the 21st Century'. Government white paper. London, HMSO, 1991

3. Bullen, P.R, Jackson, A., Taylor, P.B., 1997 'University-Industry Partnerships providing Postgraduate Programmes for the Automotive Industry'. ASME Conference, San Diego, USA.

4. Taylor, P.B., Gregory, R.D., Bullen, P.R., 1997 'Continuous Improvement Processes for Higher Education: A practical example'. SHRE Conference, Sheffield, England.

5. Harrison, R. 1992 'Employee Development'. Pub. Institute of Personnel and Development, London.

C574/023/99

Developing student capabilities and improving the local skills base – a new venture in work-based learning with the automotive industry

I DUNN, B PORTER, and **R PERKS**
School of Engineering, Coventry University, UK

KEYWORDS

Educational Development, Design Education, Student Capability, Technology Transfer, Mentoring, Work Based Learning, Reflective Practice, Action Learning.

ABSTRACT

The aim of this paper is to present the educational philosophy and educational development behind the creation of a new course entitled BSc (Hons) in Design and Technology at Coventry University. The original drivers for this course were traditional in nature; it was felt that those students from a background in a new A level in Design and Technology were not able to easily access courses in engineering in British higher education. The developments of the course have been, however, for many reasons, non-traditional.

The process that was adopted by the design team was to establish the type of applicant that may be interested by such a course. To establish the possible employment opportunities that may exist. To establish the abilities and possibilities that existed within the university and then to explore how these could be matched. The conclusion was that some of the design objectives were beyond the scope of a traditional university education, in that operational capabilities were defined for the students and these require a clear understanding of the work place environment.

The design team has attempted to involve as wide a range of people as possible, both from within the university and from industry and schools, in order to respond to the educational position that has been adopted.

All of the students will be placed with a local company for the entire course, and will spend increasing amounts of time per week in that company as the course progresses. The work-based element of the course will involve students responding to certain academic requirements through their understanding of their work situation and through guided study to contextualise the theory. These work-based elements will focus on industry in the Coventry area and will therefore be associated to the automotive industry.

Innovation also reaches to the core of the course in that the students will have an established and described role as mentors and ambassadors for the course. They will support earlier years of the course, or students in the partner schools, as well as working on the transfer of technology, and best practice, into their host company.

Groups of students will be working in action sets to explore the range of experience of contact with industry. This act of reflection is intended to develop qualities that are not necessarily subject related but are intended to enhance professional practice and planning.

In conclusion the course design team are excited by the possibilities of enhanced operational and academic capabilities of the students, the chance for technology transfer into regional small and medium sized enterprises, the closer links with the local schools and a new educational model for interested staff to explore. This paper presents the detail and process of the design phase for a new course and the theoretical underpinning for the developments.

1.0 INTRODUCTION

Coventry University School of Engineering has an excellent reputation for Engineering Design Education. We operate two courses, one in Industrial Product Design and another in Automotive Engineering Design. As a subject discipline we identified a weakness in our range of courses from two perspectives. Firstly, significant developments in the Design and Technology 'A' level in British schools mean that students' are arriving with experiences and expectations for which we were not entirely prepared. Secondly, we educate excellent product designers and excellent automotive engineering designers but we had no course that addressed a developing need for excellent design technologists. We therefore felt that we had identified a need!

When you delve further into the problem it is quite clear that you can identify a wide range of other internal and external drivers for such developments as are identified in this paper. Externally the British higher education system has recently been the subject of a major review by Ron Dearing (1). The University for Industry has recently been created, the Engineering Council is modifying its routes to accreditation, the so-called SARTOR (2) document, this course being targeted at the Incorporated Engineer level. Higher education has been developing notions of graduate capabilities (transferable skills) for a number of years, but the pressures are now quite strong to ensure that the graduate has other skills in addition to subject knowledge. Internal pressures include the Coventry University mission and strategic objectives that emphasise the importance of the university to the local economy and local industry. The City of Coventry is also involved in a Single Regeneration Budget (EU funding) bid for money to improve the local skills-base, the university is a part of that bid.

Coventry as a city has a reputation for the production of automobiles. The automotive industry is recognised as having a major impact on the development of engineering design as a subject. The combination of these factors mean that any course development that we were to carry out should pay attention to the automotive industry and preferably be in association with it.

Given that we had identified a need for a new course development and also identified a wide range of other issues that impinge, or impinged, on the design. We felt that it was appropriate that we create a design team that was able to address as many of these aspects as possible.

This paper aims to present the education thinking behind the course design and the distinct nature of the course that resulted from the design.

2.0 DEVELOPING AN EDUCATIONAL POSITION

One of the possible start points for any course design has to be 'What are possible employment prospects in this field?' In this case the answer was that we (the design team, which included members of the Engineering Design subject group, the Engineering Management subject group, and the Coventry Schools' Design and Technology Advisor) felt that positions as designers in small and medium sized companies, design detailers and managers in larger companies and design technologists in design consultancy were all possible and suitable outcomes.

Once we had established this start point we were in a position to make an attempt at the definition of the abilities (capabilities, competencies) that we felt a graduate of design and technology should have. This means establishing a course philosophy.

We felt that it was far too simplistic to expect the students to be 'competent' as designers. Our aim was to educate people who are capable of managing the design function as well as practising as designers and design technologists. Given this we felt it appropriate to explore the educational literature relating to capability and knowledge.

3.0 THE EDUCATIONAL POSITION

We based our design on four guiding principles that are, in effect, its philosophy. The first is based on the notions of student capability (3). The second is the idea of embodied and embedded knowledge (4). The third is a strong belief in experiential learning and therefore a problem-based approach as expressed diagrammatically below in Figure 1. The final of these four guiding principles is that of learning by reflection on practice (5) (6).

Figure 1: The Kolb Experiential Learning Cycle (10)

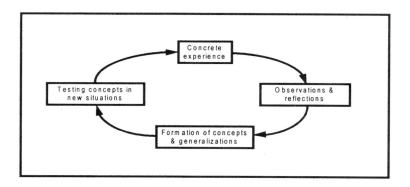

Barnett (3) uses, and in their current form partially dismisses, the ideas developed by the so-called 'competence movement', particularly associated with NVQ type qualifications. He extends these ideas to a higher education setting. The result is the definition of three types of capability, operational capability, academic capability and further life capability. Most, if not all, higher education courses can claim academic capability. Few can claim operational capability, and, although enterprise skills are a move towards further life capabilities, they are seldom embedded into the course design.

Blackler (4) presents a view of organisational knowledge that says that it is fine, indeed essential to know the facts, but the application of that knowledge in a real situation leads us to a deeper and often better understanding of the situation. He refers, therefore, to embodied and embedded knowledge, amongst others. If we attempt to link together these two pieces of work we are able to see that we are beginning, in theory, to address the issues that were presented earlier in this paper as constraints on the design of the course. Therefore, using Barnett's work we were able to construct the following exhibit, figure 2, which provides an indication of the capabilities that the course aims to help the student towards.

Given that this final iteration is felt to best describe the set of capabilities, or attributes, required of a graduate of a course in Design and Technology by both our industrial partners and particularly the automotive sector, and the academic community. This document went through a number of iterations, which eventually gave this position. A number of industrialists were consulted and made comment about the operational capabilities. The course that has been developed around them needs to have a distinct structure. To satisfy the operational capability requirements, the student needs significant exposure to industry and commerce. To satisfy the academic capability requirements the student needs a knowledge base and the tools necessary to contextualise the learning from the workplace. To develop the skills of critical reflection and continuing professional development, the student needs to be helped to understand the relationship of both the academic and the operational. This should be developed with the help of tutors, students, co-workers, and the individual, so that when faced with an unfamiliar problem in the future the student has a depth of experience on which to draw.

Figure 2: Capabilities required of a graduate of D and T

Theme	Operational capability	Academic capability	Further life capability
Epistemological	Knowledge of most modern design systems and willingness to identify best solution for situation	Knowledge of Design, Technology and Management fields at graduate level	Aware that own knowledge is limited and needs updating. Willing to question and research
Aim or focus	Solutions to immediate, detail problems	Aware of modern theory and capable of seeing relevance to practice	Willingness to debate (peers, tutors, mentors etc.) and work in teams
Communications	Able to report orally and in written form in a way appropriate to setting	Recognition of need to communicate differently in various settings	Dialogical, willing to listen and decide
Values	Recognises personal input to team and the need to co-operate	Ability to adopt disciplinary attitudes (Designer, Manager etc)	Through dialogue to a common good beyond self or company good
Boundaries	Willing to accept instruction and to instruct at a detail level	Capable of identifying and limits of knowledge and able to seek correct support	Understanding that own concepts may be flawed and willing to think beyond current experience.
Critical abilities	Capable of recognising that current systems, products or processes may be improved within resource base	Willingness to CPD	Practical understanding of colleagues requirements and consumers needs
Learning	Through experience	Individual facts, propositions, and items of information	Metalearning, reflexion
Situations	Aware that financial or power influences can result in personally unsatisfactory situations	Professionally aware of theoretical developments, and past and future trends.	Open stance in order to bring human, and value, concerns to issues
Transferability (if such a thing can exist)	Ability to apply solutions across different areas	Able to link abstract to practical knowledge and understanding	Metacritique, through a questioning attitude to develop and further own understanding
Evaluation, being able to assess impact	Economic, will it work? Working within constraints and conflicting requirements	Do I understand why it is as it is? Evaluation process.	Consensus, understanding and listening to many sides

The previous figure is constructed from the work of Ronald Barnett (3).

We feel that through a clear understanding of the needs of both the students and industry, and an understanding of the skills available within the university, the course that has been developed has a sound educational position.

The aims and objectives, assessment, teaching methods, learning styles and industrial experience must be viewed as a whole. If the teaching methods are not matched to the development of the skills and knowledge defined through the objectives and syllabus, then it is unlikely that students will receive a thorough assessment. We have defined, through detailed objectives, the quality of performance expected from the students. The objectives range from attitudes and values, through personal and mental skills, to specific skills including the process of synthesis and the application of a core knowledge syllabus.

The principle teaching approach will be problem-based, in order to satisfy the guiding principles as discussed earlier. Engineering science will be introduced as both the student and tutors determine the need for its understanding. The problems that the students address are defined by industrial need, and through regular, group, action learning activities students will gain a wide view of the relevant issues.

The assessment process will work with the teaching approach and will be formative where skills are being developed and summative where performance is being measured and a mark awarded. Self-assessment, peer assessment, and tutor assessment (academic and industrial) will be used throughout. Formal end-of-year examinations will be used in only a small number of subjects and in none of the design and design and technology activities, which will be 100% coursework assessed. It is considered that this form of assessment sits uncomfortably with the course philosophy. However, some tests will be used to assess the individual student, especially in their understanding of engineering science and design management. Importantly, assessment takes place on a continuous basis throughout the academic year.

The links with industry will be of key importance. The course aims to develop graduates who will work in all aspects of design and design management. In larger companies as developers and detailers of concepts, or as design technologists with design consultants (industrial, product, building services and architectural). In all cases taking responsibility for the process of design. The majority of industrial links will be with the automotive industry in the Coventry and Warwickshire area. This course is a partnership with the town and region. It will therefore be involved in the process of skills development in an industrial region that is defined as having a low skill base.

We have also ensured that strong links with the Coventry schools Design and Technology departments exist. Students will be required to act as mentors to pupils in these schools. This will encourage reflection on past experience for the students and will help encourage the partnership that this course can help develop in the field of design in the region.

4.0 FROM EDUCATIONAL THEORY TO EDUCATIONAL PRACTICE

Defining a course philosophy, and making it quite explicit to all concerned was a fundamental step towards creating an innovative course. A degree course in a university, however, does not exist in theory alone. As we have discussed earlier, we felt that it was necessary to create a partnership with industry and schools, in order to satisfy the learning objectives that we had established. The obvious partnership was primarily with the key employers in the region, the automotive industry. Work-based learning has been much discussed (e.g. 7, 8, 9) over the past few years, but to date little has really happened in this direction.

It is clear that initially this course is a little experimental in form and content. A significant number of questions are being asked about the way in which the work-based activity is defined and about the ability of the course to attract students. We feel that through the partnership approach that we have adopted we are able to generate enough interest from local industry. Early indications support this view. As for applications from students, we currently have demand that is significantly outstripping proposed places, and the quality of entrance qualifications is high.

4.1 Course structure

We aim to present the structure of the course as we have defined it from the preceding theoretical position. Given that we declared the Dearing report as one of the external drivers, it is clear that we have to ensure that we provide clear full-time and part-time routes.

The full-time course is of three years duration. The course will begin with 20 full-time students and then will recruit 30 to 40 students per year. All of the students are selected by both interview and academic performance. The interview aims to assess the students' suitability for a design course and for this particular course. In the first term of the first year a student will spend all of their time at the University working on modules which address both the knowledge base and the skills and attitudes required for the work place. The students during this time will attend interviews with our industrial partners. The University will have established the placements in partnership with local and regional employers and with the support of the Coventry and Warwickshire Chamber.

For the remainder of year one the student will spend approximately two days per week in the host company and approximately two days per week in the University attending classes in the more traditional sense. The students' activity in the workplace will, through negotiation between the student, the employer and the University, satisfy the company's objectives and academic learning objectives. The student will attend meetings with their peers to reflect on the weeks' activity and to ensure that the academic context of the workplace learning has been understood.

In year two the student will spend about three days per week at the company and in the final year spend three days per week at the company. The student will stay, in most instances with the same company throughout the course. A student who fails to be placed with a company through no fault of their own will be placed in one of the University's design research and consultancy units.

We are conscious that an eighteen-year-old applicant to a university course is looking for an experience of that institution that goes beyond purely the academic. For this reason, we have attempted to ensure that in years one and two the course allows time for the student to be fully integrated in student life.

For a part-time student the course will last four years, if they enter with no enhanced standing. The first stage of the course will be spread over two years because of the importance of establishing the knowledge base. The second and final stages of the course can be studied at the same pace as a full-time student. This is because of the amount of work based learning built into these stages. We feel that this is a significant blurring of the boundaries between full and part-time education. An advantage to the employer of this mode of study is that the student is subject to the normal rules of access to higher education and the associated fee regime. Therefore if the student is fee exempt the employer is gaining access to higher education and training at no fee expense.

4.2 Course content
Before addressing the actual content of the course it may be useful to define exactly what we mean by the term design technologist. In this context we are using the term to refer to a range of occupations that depend upon the size of the design activity within the company. In a small or medium sized enterprise, it is quite conceivable that design is still considered to be a 'luxury' as a separate activity, in this setting the individual would be expected to cover the whole design activity. It is also clear that in this setting the design activity may primarily consist of development work and design modification as well as process design. In a larger company where there were a number of designers, then we expect a graduate of this course to be responsible for the management of the project, the development of a design concept and the interface with the production department. Equally a graduate could be responsible for the purchase, implementation and maintenance of the computer based technologies for design. In a design consultancy setting the graduate could be responsible for the CAD activities. In all of these cases the individual would be able to act as part of a team, and would be acting as a professional engineering designer at the level of incorporated engineer.

The subject content includes:
A deep and thorough understanding of the design process, both through the workplace and academic input.
A detailed understanding of modern design technologies, CATIA, AutoCAD, Knowledge Based Engineering, Rapid Prototyping and so on.
Detailed knowledge of current design tools and techniques, FMEA, QFD, DfX etc.
An understanding of the uses of Information Technology for communication.
A clear understanding of design management techniques, project planning and management, finance, marketing, total quality management, operations management, innovation management.
A good understanding of manufacturing processes, technologies and systems.
A clear overview of engineering science, mechanical, electrical and mathematical.

All of this must be supported by the ability to keep a professional portfolio of work, a reflective journal and also the ability to communicate effectively at all levels of activity.

We are conscious that the student is experiencing one workplace, and that the workplace mentor is not trained to assess the student. For this reason the student will be assessed, in the summative sense, by academic tutors, on the evidence presented in their design portfolio and reflective journal. This assessment will be supported by the evidence collected through a significant number of industrial visits and letters of comment from the industrial mentor.

4.3 Local skills enhancement
One of the stated aims is that the course will act as a means by which initially local, but later regional, industry can gain access to modern tools and techniques. The students on the course

are going to be working with local companies as employees/advisors. An element of their activity is to undertake a skills/technology needs analysis for the company. Having performed this it is expected that they will then act as trainers within the workplace. The benefits to the student are clear, teaching someone how to do something in as short a time as possible requires a clear understanding of the activity from the trainer. The company benefits from the experience of the student as well as access to information through the University.

5.0 CONCLUSIONS

We have presented the development of programme of study at Coventry University. We can not claim that this is the only course that includes work-based study. We can not claim that we will not be required to refine and improve the structure and content as we learn from our mistakes. We do claim however that we have developed a thoughtful and sound educational position from which we have produced an innovative programme. We can claim to offer students the chance to learn in a new and challenging manner, that will require great commitment from them but will in turn deliver to them a depth of understanding of their subject that would have been impossible by a traditional approach. We can claim that we are creating a partnership with our region, opening up educational provision to individuals and companies that may not have accessed it previously. This partnership exists primarily with industry through the automotive sector and with the region through our local schools.

We have considered the current state of design at degree level and established that there was a need for a new type of design engineer. A person who has a clear understanding of design and is equipped with the tools and techniques that are currently being used and developed, but also with the ability to keep up-to-date.

In final conclusion, through a good understanding of educational theory and an awareness of the current trends in higher education and a good understanding and experience of the subject, we have been able to assemble a course that may prove to be a model for other engineering disciplines.

6.0 REFERENCES

(1) The National Committee of Inquiry into Higher Education (1997) *Higher Education in the Learning Society*, HMSO

(2) SARTOR 97 (1997), The Engineering Council, London

(3) Barnett, R (1994) *The Limits of Competence: Knowledge, Higher Education and Society*, OU Press

(4) Blackler, F (1995) Knowledge, Knowledge work and Organizations: An Overview and Interpretation, *Organization Studies*, v.16 no.6 pp 1021-1046

(5) McGill, I and Beaty, E (1995) *Action Learning: A Guide for Professional Management and Educational Development*, London: Kogan Page

(6) Schön, DA (1991) *The Reflective Practitioner: How Professionals Think in Action*, Arena

(7) Duckenfield, M and Stirner P (1992) *Learning Through Work*, Employment Department Report on Higher Education Developments, London

(8) Portwood, D (1993) Work based learning: linking academic and vocational qualifications, *Journal of Further and Higher Education*, Vol. 17 No. 3 pp. 61-69

(9) Kinman, R and Kinman, G (1997) Work-based learning on trial, *Industry and Higher Education*, Vol. 11 No. 5 pp. 314-321

(10) Kolb, D (1988) *Experiential Learning: Experience as the source of Learning Development*, Prentice Hall

Industrial PhD

C574/012/99

Global Product Development – an integral part of an engineering doctorate in automotive engineering management

P B TAYLOR and **P R BULLEN**
Department of Aerospace, Civil, and Mechanical Engineering, University of Hertfordshire, Hatfield, UK
M D COOK
Aspirations Consultancy, Bedford, UK

ABSTRACT

The University of Hertfordshire has designed an Engineering Doctorate (EngD) scheme specifically for the Automotive Industry. A significant, and innovative, phase of the EngD will be an intensive two week course in Global Product Development followed by a semester long Global Project which will bring together Research Engineers and Applied Doctorate students from all around the world. The goals of this course are to rapidly build an understanding of the interplay of culture, engineering and business in global management of the manufacturing enterprise. This paper seeks to place this project in the wider context of the EngD qualification and to share the experiences and approaches used in specifically developing and delivering this element of the EngD, which has challenged both the developers and the Research Engineers into global thinking, teamwork and multi-cultural management.

1. BACKGROUND

"The only strategy for a nation seeking to maintain and enhance a high standard of living lies in concentration on advanced products and services, a high level of innovation, challenging and constantly improving standards of achievement and competitiveness, based on a highly educated well trained and adaptable workforce" Sir Ron Dearing 1996

The Engineering Doctorate (EngD) was introduced in 1992, as a five year pilot scheme, following the recommendations of the Parnaby Report. Over four years, it provides engineers with the business and technical competencies by applying new knowledge to industrial relevant doctoral research, employing the skills gained from an intensive programme of taught coursework. (1) Students on this scheme are known as 'Research Engineers' (RE's).

The University of Hertfordshire has designed and developed an Engineering Doctorate scheme specifically for the Automotive Industry. This scheme is designed to speed up the development of senior engineering managers and specialist automotive engineers, whilst maintaining the rigorous training of the traditional PhD programme. Two parallel schemes are available – Automotive Engineering and Automotive Engineering Management. A key part of both schemes will be looking at the issues that influence product development on a global scale. The scheme will link to a similar project in Wayne State University, Detroit with a view to developing a innovative curriculum in global engineering and manufacturing management. This paper seeks to position the importance of a significant phase of the EngD which is an intensive two week course in Global Product Development followed by a semester long Global Project which will bring together Research Engineers and Applied Doctorate students from around the world.

2. AIMS OF THE ENGINEERING DOCTORATE

Both the programmes discussed here, are designed to produce engineering managers and specialist automotive engineers who are experts in key areas of technology and capable of working at the leading edge in their field. They will be leaders with an innovative approach to managing change, complexity and diversity, yet also capable of rigorous analysis.

These engineering managers will meet the needs of the industry by maintaining and developing their competitive advantage, choosing and developing key areas of technology and will strive to out-perform international best practice.

In the past these individuals have developed partly through experience and partly through education. The PhD, traditionally associated with developing technical experts, has tended to have a narrow focus. The EngD programme is designed to overcome this problem, speeding up the process of developing senior engineering managers and specialist automotive engineers, meeting the needs of the individual and of industry, whilst maintaining and enhancing the rigorous training of the PhD programme. It is designed to suit the industry's needs by incorporating the following features. It can be Company or University based, it seeks to develop expertise, rigorous analysis and evaluation, innovative thinking and application to industrial problems. Whilst broadening knowledge to increase the understanding of the overall context, this EngD will develop and integrate management, personal and technical skills.

The programme, as with many other EngD schemes, is integrated with an existing MSc. In this case the MSc in Automotive Engineering Design, Manufacture and Management, an Integrated Graduate Development Scheme (IGDS). This scheme is part-time and industry-focussed and the individual subject modules will provide underpinning knowledge to support the programme of study for the Engineering Doctorate. In this way the training is externally accredited as high quality and at postgraduate level. (1)

The specific aims of the Automotive Engineering Doctorate degrees are to increase and develop skills in key areas of technology, management theory and practice. This will include education in the processes and infrastructure of the automotive engineering business, critical analysis and research skills, as well as dealing with complex issues including managing innovation, change, diversity and very importantly, a multidisciplinary globally dispersed team. It is designed to enable the Research Engineer to make a significant contribution to the enterprise within which the programme of study is carried out.

3. STRUCTURE AND OUTLINE CONTENT

The structure of the programme is indicated in the diagram below -

Phase 0	Phase 1	Phase 2	Phase 3
Courses and assignments	Courses and assignments	Courses and assignments	Individual Project
Project 1 (Individual)	Extension to Project 1 or new project		
	Preparation for and introduction to global team project (optional)	Preparation for global team project (optional)	Preparation of thesis and portfolio
	Global team project or individual project	Global team project (optional) or individual project	
BEng/MEng	MSc	MSc(RES)	EngD

Figure 1- Outline Structure of the EngD programme

4. KEY OVERALL FEATURES

- The EngD is primarily an engineering qualification where the assessment of the project work is through the same University route as for other Doctoral degrees and therefore quality is assured in the same way.
- Entry to phase 0 requires a good engineering first degree
- The outcome of phase 0 could lead to an MSc on completion of IGDS requirements
- Entry to phase 1 requires an engineering first degree and a Masters in engineering or associated discipline (e.g. MSc, MBA)
- Phase 1 consists of further courses from the programme of supporting studies, individual project work, preparation and participation in a global team project, or individual project. Some courses are compulsory and depend on entry qualification and experience
- Phase 2 includes similar activities to phase 1, involving an optional second global team project. Successful completion of phase 2 could lead to an MSc (by research) or continuation to an EngD
- Phase 3 will concentrate on a final project and preparation of the final thesis and portfolio of work. This requires a report for each project together with a critical appraisal of the research programme submitted as a thesis. Each project will form part of an overall theme agreed at the start of the programme.

- The outcome of Phase 3 is an EngD
- Courses and assignments will provide a 'toolkit' to support the projects, the development of the individual and the understanding of the wider context and will include subjects within the themes of Automotive Engineering Design, Manufacture and Management and Research methods. These will be supported by the new technologies for communications, teaching and learning and student support. Courses will take up 20 – 30% of the programme as recommended by the Engineering Doctorate Review Panel. (1)

5. KEY FEATURES OF THE PROJECTS

The programme includes a number of 'projects' (minimum 2, maximum 4) linked through a common theme which are assessed by a thesis and portfolio of work. The projects should enable the demonstration of engineering management, engineering innovation and application of theoretical concepts combined with rigorous analysis and evaluation of outcomes whilst making a significant contribution to the performance of the company. The projects will generally be company based, linked closely with the 'day to day' work of the Research Engineer. Implementation of aspects of the projects can be carried out by others, but must be managed by the Research Engineer. The projects should also allow the Research Engineer to demonstrate the ability to both manage and participate in a globally dispersed team (particularly for Automotive Engineering Management).

6. THE GLOBAL DIMENSION

The major automotive manufacturing companies now operate on a global basis producing automotive products which benefit from global economies of scale and which are sucessful in diverse markets. This requires individuals with the ability to manage multinational teams, that are globally dispersed, and who understand and can deal with the increased complexity of product development manufacture and management. The Automotive Engineering management (AEM) EngD programme is designed to provide this global dimension through programme design and close collaboration with Wayne State University, Detroit, USA (and other partner Universities around the world). Included in the programme of work will be a project on 'Globalisation' and its impacts. This project is a Global Team Project, involving working with a multi-disciplinary, globally dispersed team and will start with a two-week intensive residential workshop followed by a semester long project whilst the Research Engineer is back at work.

The Global Team Project focuses on a global case study and has been developed, and will be delivered, by a multi-national, multi-disciplinary team from engineering, anthropology and business from both Universities and will include business leaders from the Automotive and related industries.

7. DEVELOPMENT OF THE GLOBAL TEAM PROJECT

It was decided that the development team should utilise the same tools for developing the project, as the Research Engineers would be expected to use to execute the project. This would lead to learning at various levels and would hopefully identify 'best-practice' in using the new communication technologies.

The core team of staff; four from each University held teleconferences using a Pictel videoconferencing set-up, to discuss and plan the programme. Other communication channels e.g. e-mail, fax and telephone were also used. It was quickly recognised that a strict discipline and protocol was necessary to gain maximum benefit from each teleconference. It was very important to develop a level of trust with all team members especially between the two universities. Prior to each teleconference, an agenda was prepared and circulated for agreement and each team member took responsibility for preparing for the meeting and carrying out any action point for which they had volunteered to take the lead. The mix of engineering, anthropology and business allowed a variety of perspectives and a wealth of experience to be brought to the discussion and planning. Members of the Wayne State team had already developed and delivered some areas of business, management and product development with a global focus and there was a significant level of experience of working with and developing multi-cultural teams.

To further 'prove out' this concept a pilot project was undertaken using a case study which a team of MSc. students worked through together, half of the team in the UK and half in the US. This project took the form of two formal 2-hour videoconferences, numerous phone calls, e-mails and faxes and a final review of learning outcomes with the entire team. From this exercise it was found that the technology used was reliable and appropriate for this application, the students put in a lot more time and effort than they would have for a locally based team project, but equally felt that they had learned some important cultural lessons. A set of guidelines were produced from this pilot project, which if followed allow maximum communication and learning opportunities. These include – keeping the formal sessions very structured with aims clearly defined, having a detailed agenda with times given, appointing a Chair and issuing an attendance list before the session to introduce the group and their skills (a seating plan also helps!). The students pointed out that barriers to communication could arise more easily at a distance, so it is important to discuss cultural and communication issues openly, and preferably in advance of the large group formal sessions. The success of this pilot project has directly contributed to the development of the Global Team Project and lessons learnt are being written in to the course structure. (2)

7.1 The Two Week Intensive Workshop

The goals of the experience are to rapidly build an understanding of the interplay of culture, engineering and business in global management of the manufacturing enterprise whilst applying knowledge to a global engineering case study within a framework of culture, recognising the business environment in the host region. It is intended that the hosting of the programme will alternate each year, between the University of Hertfordshire and Wayne State University.

In pursuit of these goals the Research Engineers will actively participate in establishing, and developing global team relationships through working in a globally dispersed team. They will acquire global teaming skills including the use of communications technology management and they will be introduced to tools, methods and approaches that can be used to support these activities. A very important aspect of this workshop is the Research Engineer's definition and understanding of their role in the semester long global multicultural team project.

A multidisciplinary team from engineering, anthropology and business will teach the group of Research Engineers and Applied Doctorate students. A variety of learning strategies will be employed, the main one being student-centred experiential learning with ample opportunities for reflection, negotiation and sharing. This is complemented by formal teaching inputs and a wide variety of different teaching and learning methods including: discussions, role-play, case study, simulation, self-directed reading, visiting speakers, assessed group tasks, presentations, and project work. Emphasis will be placed on group and shared activities as the globally dispersed members come to form an established cohesive group from which they can truly value and benefit from each other's experience.

During the two-week workshop, Research Engineers will experience a variety of topics including: aspects of globalisation and global manufacturing and management, with strong emphasis on cultural content, marketing and product development. There will be a focus on the business environment of the host region including economic, marketing and geographical influences. The Research Engineers will be introduced to tools, methods and approaches used within global engineering teams including communications technology. There will also be an engineering case study to be undertaken as a multi-cultural global team with the findings presented at the end of the course. This will allow the Research Engineers to participate in a multi-cultural team and gain experience using the various frameworks, techniques and models introduced during the workshop. Definition and initial planning of the semester long project together with team building in a multi-cultural global environment including cultural consideration, team roles, team dynamics, communication, management and leadership will provide greater understanding of individual learning and problem solving styles whilst valuing difference.

During the two week period students will experience three sessions each day, morning, afternoon and evening. The workshop is divided into sessions as follows:

7.1.1 *Introduction (1-day)*
All Research Engineers and tutors will come together to introduce each other as well as the programme and to discuss expectations and experiences and to determine individual and group objectives.

7.1.2 *Defining the Environment (1-day)*
This covers the challenges, threats and opportunities that globalisation in general offers with specific reference to the automotive industry.

7.1.3 *Setting the Stage for Global teaming (2-days)*
Topics covered in this section will focus on cultural content, global marketing and product development in a global arena. Lectures, case studies and visiting speakers will feature in this session.

7.1.4 *Small Project (5.5-days)*
The small project which will be scoped and developed during the two week period will allow Research Engineers to utilise techniques and methods gained during the programme whilst developing relationships and understanding of each other and the team. Deliverables from the Small project will be a Project Plan Proposal to include 3 options/alternatives, estimates of resources required including money, people, time, systems, equipment etc. with financial implications. Then strategies and plans for the project development will include: Marketing,

Finance, Global and Local Communication, Global and Local business environments, IT including Data Warehouse, Knowledge Management and whatever else is considered appropriate by the project team. The Project Team will present a review of their performance, to include how they managed /monitored the project, team/individual responsibilities, progress versus plan, problems encountered and how these were overcome, tools, methods and approaches used. What have they learned from the experience, and what would they do differently next time? During the Small Project the Tutors will engage in "participant observation" of the Research Engineers as a means of gathering information about how they do their work, what issues they confront etc. and this will be fed back to the students.

7.1.5 *Scoping the Semester-long Global Project (0.5 day)*
The final phase of the two week intensive programme will be to scope and provide an initial plan for the semester long project. Each Research Engineer will provide a report which should include the aims and objectives of the project, the report is then personalised by each student documenting their part in the project, and what their individual tasks/actions with timescales and dates. They are also required to state what role they play in the team, and how the project will be integrated across the group.

These aspects of the two-week workshop will be further supported with team building activities, case studies and visiting speakers from the automotive and other global industries.

7.2 The Semester-long Global Project
The Research Engineers will then carry out the project from their workplace during the following semester, using video conferencing and Internet technologies to replicate the challenges of the working environment. Following the scoping activity carried out during the two-week programme, the teams will be expected to provide protocols and plans for their methods and frequency of communicating and reporting to each other and to Tutors.

Regular teleconferences will be held and a formal Interim Presentation in the form of a videoconference team presentation will be required for assessment by tutors. This presentation will take place about 2/3rds through the semester long project, stating what has been achieved, how they are meeting their goals, problems encountered and solutions adopted etc. The Global Team Project will culminate with an assessed team presentation accompanied by the final report – each Research Engineer will produce his or her own report.
Half of the report focuses on the team, (management, culture, learning experiences etc.) how the team worked together, or not, recommendations for improvement, etc. the remainder is about the individual's contributions and the technical input to the project.

Key to the success of this Global Project, is the establishment and discipline of team working protocols with strong emphasis on the media and frequency of communication. This is a key requirement for both the tutor teams and Research Engineer teams. With the pace of change and innovation within telecommunications and associated technology, the distances and time differences will be managed more easily as has been proven through the development of the programme.

8. CONCLUSIONS

The Engineering Doctorate programme is seen as a major development in increasing the educational level of senior engineering managers whilst strengthening their 'on the job' performance. The scheme provides a background of underpinning study and academic rigour for industrial based activities whilst encouraging contributions to the 'bottom line' performance of the delegates sponsoring organisation.

A major innovation in the development of the of the Engineering Doctorate is the development of the Global Team Project which strives to give delegates experience of working in a global team, on a 'real' business problem in a 'real' global environment with 'real' timescales and using the latest communication technologies. This teamworking will provide a valuable forum for sharing ideas, networking and learning across the whole automotive industry.

The development and delivery of this course, has challenged both the developers and the delegates into global thinking, teamwork and multi-cultural management. Year on year more understanding of the global environment will be gained and more experience related to global engineering management will be added to the programme. Additionally, new developments will take place building on the experience that the team gains in running each programme and the feedback from the Research Engineers.

REFERENCES

1. Engineering Doctorate Review Panel, EngD Review Document (1998), EPSRC, London
2. Taylor, P., 'Global Team Projects – A Pilot Initiative Undertaken to Establish Feasibility'. SHRE Conference 'Tomorrow's World: The Globalisation of Higher Education' University of Lancaster 1998

FURTHER INFORMATION

For general information on the Engineering Doctorate scheme access the EPSRC web site at http://www.epsrc.ac.uk/EPSRCWEB/MAIN/TRAINING/CONSID/INTRO/INTRO.asp?main.htm
This gives access to a full list of EPSRC supported EngD schemes and detailed documentation on the requirements of the scheme, with case studies of some recent graduates.

Authors' Index

B

Barlow, N 21–30
Barton, D C 33–42
Brooks, P C 33–42
Bullen, P R 45–54, 77–88, 91–100,
.... 113–120

C

Chatterton, P F 21–30, 77–88
Cook, M D 113–120
Crolla, D A 33–42

D

Day, A J 55–66
Deakin, A J 33–42
Dunn, I 101–110

G

Glover, A 21–30
Gordon, T J 67–76

H

Haddleton, F 77–88
Harding, R S F 55–66

J

Jones, M 21–30

L

Lyons, A C 21–30

M

Malalasekera, A 67–76
Mortimer, K W 55–66
Mughal, H 45–54
Mulryan, J A 91–100

O

Oxtoby, B 21–30

P

Perks, R 101–110
Porter, B 101–110
Priest, M 33–42

R

Russell, M F 3–20

T

Taylor, P B 45–54, 77–88, 91–100,
.... 113–120

W

Walsh, S J 67–76

Y

Young, M 77–88